ESSENTIALS OF ENZYMOLOGY

ESSENTIALS OF ENZYMOLOGY

By

GAYATHRI G R
K. G. SREEJALEKSHMI

KNOWLEDGE BAKERS

ISBN: 978-93-49964-16-7

Published by : **Knowledge Bakers**
Pune, India
info@knowledgebakers.com

PREFACE

Enzymes are fundamental to life, acting as catalysts that accelerate the vast array of biochemical reactions necessary for maintaining cellular functions. From digestion and metabolism to DNA replication and signal transduction, enzymes are involved in virtually every aspect of biological activity. The study of enzymes, known as enzymology, plays a pivotal role in biochemistry, molecular biology, and biotechnology, providing invaluable insights into the mechanisms of life processes. Understanding enzyme structure, function, kinetics, and regulation is essential for advancing various fields, including medicine, agriculture, and environmental science.

This book, titled Essentials of Enzymology, provides a comprehensive introduction to the principles and applications of enzymology. It is designed for students, researchers, and professionals in biochemistry, molecular biology, pharmacology, and related fields. Whether used as a textbook for advanced coursework or as a reference for professionals, this volume covers the essential concepts and practical knowledge required to understand enzyme function and its broader implications.

The book begins with a foundational overview of enzyme structure and function, detailing the role of enzymes as biological catalysts. It explores the principles of enzyme kinetics, including enzyme-substrate interactions, reaction mechanisms, and factors that influence enzyme activity. Additionally, the book examines enzyme regulation, highlighting how cells control enzymatic activity in response to internal and external signals, which is crucial for maintaining cellular homeostasis.

Essentials of Enzymology also discusses the practical applications of enzymology in a variety of fields. It covers the role of enzymes in medicine, including their use in diagnostics and therapeutics, as well as their applications in industrial processes such as biofuel production, food processing, and pharmaceuticals. The book provides insights into enzyme engineering, exploring how recombinant DNA technology and directed evolution are used to create enzymes with enhanced properties for specific applications.

Throughout the text, case studies, real-world examples, and cutting-edge research illustrate the diverse and rapidly evolving applications of enzymology. The book emphasizes the interdisciplinary nature of the field, encouraging readers to explore the connections between enzymology and other areas of scientific inquiry.

It is our hope that Essentials of Enzymology will inspire a deeper understanding of the crucial role enzymes play in living systems and offer a valuable resource for anyone seeking to explore the fascinating world of enzymology. As research in this area continues to advance, the potential for enzymatic solutions to some of the world's most pressing challenges—such as disease treatment, environmental sustainability, and food security—remains vast and promising.

Gayathri G R

K. G. Sreejalekshmi

CONTENTS

Chapter 1

A BRIEF HISTORY OF ENZYMOLOGY

Life depends on a well-orchestrated series of chemical reactions. Many of these reactions, however, proceed too slowly on their own to sustain life. Hence nature has designed catalysts, which we now refer to as *enzymes*, to greatly accelerate the rates of these chemical reactions. The catalytic power of enzymes facilitates life processes in essentially all life-forms from viruses to man. Many enzymes retain their catalytic potential after extraction from the living organism, and it did not take long for mankind to recognize and exploit the catalytic power of enzyme for commercial purposes. In fact, the earliest known references to enzymes are from ancient texts dealing with the manufacture of cheeses, breads, and alcoholic beverages, and for the tenderizing of meats.

Today enzymes continue to play key roles in many food and beverage manufacturing processes and are ingredients in numerous consumer products, such as laundry detergents (which dissolve protein-based stains with the help of proteolytic enzymes). Enzymes are also of fundamental interest in the health sciences, since many disease processes can be linked to the aberrant activities of one or a few enzymes. Hence, much of modern pharmaceutical research is based on the search for potent and specific inhibitors of these enzymes. The study of enzymes and the action of enzymes has thus fascinated scientists since the dawn of history, not only to satisfy erudite interest but also because of the utility of such knowledge for many practical needs of society. This brief chapter sets the stage for our studies of these remarkable catalysts by providing a historic background of the development of enzymology as a science.

We shall see that while enzymes are today the focus of basic academic research, much of the early history of enzymology is linked to the practical application of enzyme activity in industry.

1.1 ENZYMES IN ANTIQUITY

The oldest known reference to the commercial use of enzymes comes from a description of wine making in the Codex of Hammurabi (ancient Babylon, circa 2100 ス.ズ.). The use of microorganisms as enzyme sources for fermentation was widespread among ancient people. References to these processes can be found in writings not only from Babylon but also from the early civilizationsof Rome, Greece, Egypt, China, India. Ancient texts also contain a number of references to the related process of vinegar production, which is based on the enzymatic conversion of alcohol to acetic acid. Vinegar, it appears, was a common staple of ancient life, being used not only for food storage and preparation but also for medicinal purposes.

Dairy products were another important food source in ancient societies. Because in those days fresh milk could not be stored for any reasonable length of time, the conversion of milk to cheese became a vital part of food production, making it possible for the farmer to bring his product to distant markets in an acceptable form. Cheese is prepared by curdling milk via the action of any of a number of enzymes. The substances most commonly used for this purpose in ancient times were ficin, obtained as an extract from fig trees, and rennin, as rennet, an extract of the lining of the fourth stomach of a multiple-stomach animal, such as a cow. A reference to the enzymatic activity of ficin can, in fact, be found in Homer's classic, the Iliad:

As the juice of the fig tree curdles milk, and thickens it in a moment though it be liquid, even so instantly did Paee¨on cure fierce Mars.

The philosopher Aristotle likewise wrote several times about the process of milk curdling and offered the following hypothesis for the action of rennet:

Rennet is a sort of milk; it is formed in the stomach of young animals while still being suckled. Rennet is thus milk which contains fire, which comes from the heat of the animal while the milk is undergoing concoction.

Another food staple throughout the ages is bread. The leavening of bread by yeast, which results from the enzymatic production of carbon dioxide, was well known and widely used in ancient times. The importance of this process to ancient society can hardly be overstated.

Meat tenderizing is another enzyme-based process that has been used since antiquity. Inhabitants of many Pacific islands have known for centuries that the juice of the papaya fruit will soften even the toughest meats. The active enzyme in this plant extract is a protease known as papain, which is used even today in commercial meat tenderizers. When the British Navy began exploringthe Pacific islands in the 1700s, they encountered the use of the papaya fruit as a meat tenderizer and as a treatment for ringworm. Reports of these native uses of the papaya sparked a great deal of interest in eighteenth-century Europe, and may, in part, have led to some of the more systematic studies of digestive enzymes that ensued soon after.

1.2 EARLY ENZYMOLOGY

While the ancients made much practical use of enzymatic activity, these early applications were based purely on empirical observations and folklore, rather than any systematic studies or appreciation for the chemical basis of the processes being utilized. In the eighteenth and nineteenth centuries scientists began to study the actions of enzymes in a more systematic fashion. The process of digestion seems to have been a popular subject of investigation during the years of the enlightenment. Wondering how predatory birds manage to digest meat without a gizzard, the famous French scientist Re´aumur (1683—1757) performed some of the earliest studies on the digestion of buzzards. Re´aumur designed a metal tube with a wire mesh at one end that would hold a small piece of meat immobilized, to protect it from the physical action of the stomach tissue. He found that when a tube containing meat was inserted into the stomach of a buzzard, the meat was digested within 24 hours.

Thus he concluded that digestion must be a chemical rather than a merely physical process, since the meat in the tube had been digested by contact with the gastric juices (or, as he referred to them, "a solvent"). He tried the same experiment with a piece of bone and with a piece of a plant. He found that while meat was digested, and the bone was greatly softened by the action of the gastric juices, the plant material was impervious to the "solvent"; this was probably the first experimental demonstration of enzyme specificity.

Re´aumur's work was expanded by Spallanzani (1729—1799), who showed that the digestion of meat encased in a metal tube took place in the stomachs of a wide variety of animals, including humans. Using his own gastric juices, Spallanzani was able to perform digestion experiments on pieces of meat in vitro (in the laboratory). These experiments illustrated some critical features of the active ingredient of gastric juices: by means of a control experiment in which meat treated with an equal volume of water did not undergo digestion Spallanzani demonstrated the presence of a specific active ingredient in gastric juices. He also showed that the process of digestion is temperature dependent, and that the time required for digestion is related to the amount of gastric juices applied to the meat. Finally, he demonstrated that the active ingredient in gastric juices is unstable outside the body; that is, its ability to digest meat wanes with storage time.

Today we recognize all the foregoing properties as common features of enzymatic reactions, but in Spallanzani's day these were novel and exciting findings. The same time period saw the discovery of enzyme activities in a large number of other biological systems. For example, a peroxidase from the horseradish was described, and the action of _-amylase in grain was observed. These early observations all pertained to materials—crude extract from plants or animals—that contained enzymatic activity..

During the latter part of the nineteenth century scientists began to attempt fractionations of these extracts to obtain the active ingredients in pure form.

For example, in 1897 Bertrand partially purified the enzyme laccase from tree sap, and Buchner, using the "pressed juice" from rehydrated dried yeast, demonstrated that alcoholic fermentation could be performed in the absence of living yeast cells. Buchner's report contained the interesting observation that the activity of the pressed juice diminished within 5 days of storage at ice temperatures. However, if the juice was supplemented with cane sugar, the activity remained intact for up to 2 weeks in the ice box. This is probably the first report of a now well-known phenomenon—the stabilization of enzymes by substrate. It was also during this period that Ku¨ hne, studying catalysis in yeast extracts, first coined the term "enzyme" (the word derives from the medieval Greek word *enzymos*, which relates to the process of leavening bread).

1.3 THE DEVELOPMENT OF MECHANISTIC ENZYMOLOGY

As enzymes became available in pure, or partially pure forms, scientists' attention turned to obtaining a better understanding of the details of the reaction mechanisms catalyzed by enzymes. The concept that enzymes form complexes with their substrate molecules was first articulated in the late nineteenth century. It is during this time period that Emil Fischer proposed the "lock and key" model for the stereochemical relationship between enzymes and their substrates; this model emerged as a result of a large body of experimental data on the stereospecificity of enzyme reactions. In the early twentieth century, experimental evidence for the formation of an enzyme—substrate complex as a reaction intermediate was reported. One of the earliest of these studies, reported by Brown in 1902, focused on the velocity of enzyme-catalyzed reactions. Brown made the insightful observation that unlike simple diffusionlimited chemical reactions, in enzyme-catalyzed reactions "it is quite conceivable . . . that the time elapsing during molecular union and transformation may be sufficiently prolonged to influence the general course of the action." Brown then went on to summarize the available data that supported the concept of formation of an enzyme—substrate complex:

There is reason to believe that during inversion of cane sugar by invertase the sugar combines with the enzyme previous to inversion. C. O'Sullivan and Tompson . . . have shown that the activity of invertase in the presence of cane sugar survives a temperature which completely destroys it if cane sugar is not present, and regard this as indicating the existence of a combination of the enzyme and sugar molecules. Wurtz [1880] has shown that papain appears to form an insoluble compound with fibrin previous to hydrolysis. Moreover, the more recent conception of E. Fischer with regard to enzyme configuration and action, also implies some form of combination of enzyme and reacting substrate.

Observations like these set the stage for the derivation of enzyme rate equations, by mathematically modeling enzyme kinetics with the explicit involvement of an intermediate enzyme–substrate complex. In 1903 Victor Henri published the first successful mathematical model for describing enzyme kinetics. In 1913, in a much more widely read paper, Michaelis and Menten expanded on the earlier work of Henri and rederived the enzyme rate equation that today bears their names. The Michaelis–Menten equation, or more correctly the Henri–Michaelis–Menten equation, is a cornerstone of much of the modern analysis of enzyme reaction mechanisms.

The question of how enzymes accelerate the rates of chemical reactions puzzled scientists until the development of transition state theory in the first half of the twentieth century. In 1948 the famous physical chemist Linus Pauling suggested that enzymatic rate enhancement was achieved by stabilization of the transition state of the chemical reaction by interaction with the enzyme active site. This hypothesis, which was widely accepted, is supported by the experimental observation that enzymes bind very tightly to molecules designed to mimic the structure of the transition state of the catalyzed reaction.

In the 1950s and 1960s scientists reexamined the question of how enzymes achieve substrate specificity in light of the need for transition state stabilization by the enzyme active site. New hypotheses, such as the "induced fit" model of Koshland emerged at this time to help rationalize the competing needs of substrate binding affinity and reaction rate enhancement by enzymes. During this time period, scientists struggled to understand the observation that metabolic enzyme activities can be regulated by small molecules other than the substrates or direct products of an enzyme. Studies showed that indirect interactions between distinct binding sites within an enzyme molecule could occur, even though these binding sites were quite distant from one another. In 1965 Monod, Wyman, and Changeux developed the theory of allosteric transitions to explain these observations. Thanks in large part to this landmark paper, we now know that many enzymes, and nonenzymatic ligand binding proteins, display allosteric regulation

1.4 STUDIES OF ENZYME STRUCTURE

One of the tenets of modern enzymology is that catalysis is intimately related to the molecular interactions that take place between a substrate molecule and components of the enzyme molecule, the exact nature and sequence of these interactions defining per se the catalytic mechanism. Hence, the application of physical methods to elucidate the structures of enzymes has had a rich history and continues to be of paramount importance today. Spectroscopic methods, x-ray crystallography, and more recently, multidimensional NMR methods have all provided a wealth of structural insights on which theories of enzyme mechanisms have been built. In the early part of the twentieth century, x-ray crystallography became the premier method for solving the structures of small molecules. In 1926 James Sumner published the first crystallization of an enzyme, urease (Figure 1.1). Sumner's paper was a landmark contribution, not only because it portended the successful application of x-ray diffraction for solving enzyme structures, but also because a detailed analysis allowed Sumner to show unequivocally that the crystals were composed of protein and that their dissolution in solvent led to enzymatic activity. These observations were very important to the development of the science of enzymology because they firmly established the protein composition of enzymes, a view that had not been widely accepted by Sumner's contemporaries.

Sumner's crystallization of urease opened a floodgate and was quickly followed by reports of numerous other enzyme crystals. Within 20 years of Sumner's first paper more than 130 enzyme crystals had been documented. It was not, however, until the late 1950s that protein structures began to be solved through x-ray crystallography. In 1957 Kendrew became the first to deduce from x-ray diffraction the entire three-dimensional structure of a protein, myoglobin. Soon after, the crystal structures of many proteins, including enzymes, were solved by these methods. Today, the structural insights gained from x-ray crystallography and multidimensional NMR studies are commonly used to elucidate the mechanistic details of enzyme catalysis, and to design new ligands (substrate and inhibitor molecules) to bind at specific sites within the enzyme molecule.

The deduction of three-dimensional structures from x-ray diffraction or NMR methods depends on knowledge of the arrangement of amino acids along the polypeptide chain of the protein; this arrangement is known as the amino acid sequence. To determine the amino acid sequence of a protein, the component amino acids must be hydrolyzed in a sequential fashion from the polypeptide chain and identified by chemical or chromatographic analysis.

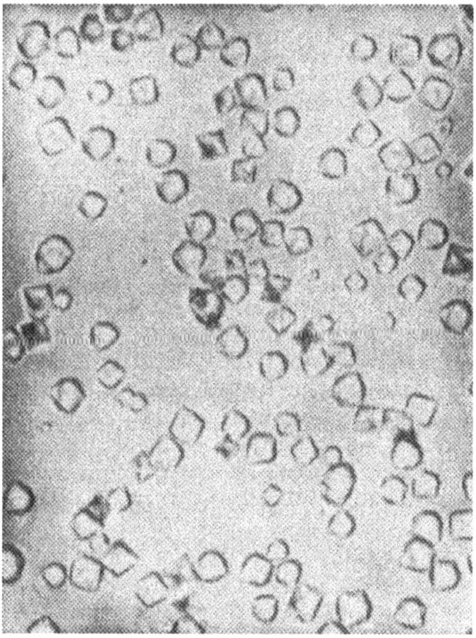

Figure 1.1 Photomicrograph of urease crystals (728- magnification), the first reported crystals of an enzyme.

1.5 ENZYMOLOGY TODAY

Fundamental questions still remain regarding the detailed mechanisms of enzyme activity and its relationship to enzyme structure. The two most powerful tools that have been brought to bear on these questions in modern times are the continued development and use of biophysical probes of protein structure, and the application of molecular biological methods to enzymology.

X-ray crystallography continues to be used routinely to solve the structures of enzymes and of enzyme—ligand complexes. In addition, new NMR methods and magnetization transfer methods make possible the assessment of the three-dimensional structures of small enzymes in solution, and the structure of ligands bound to enzymes, respectively.

The application of Laue diffraction with synchrotron radiation sources holds the promise of allowing scientists to determine the structures of reaction intermediates during enzyme turnover, hence to develop detailed pictures of the individual steps in enzyme catalysis. Other biophysical methods, such as optical (e.g., circular dichroism, UV—visible, fluorescence) and vibrational (e.g., infrared, Raman) spectroscopies, have likewise been applied to questions of enzyme structure and reactivity in solution. Technical advances in many of these spectroscopic methods have made them extremely powerful and accessible tools for the enzymologist. Furthermore, the tools of molecular biology have allowed scientists to clone and express enzymes in foreign host organisms with great efficiency. Enzymes that had never before been isolated have been identified and characterized by molecular cloning. Overexpression of enzymes in prokaryotic hosts has allowed the purification and characterization of enzymes that are available only in minute amounts from their natural sources. This has been a tremendous advance for protein science in general. The tools of molecular biology also allow investigators to manipulate the amino acid sequence of an enzyme at will. The use of site-directed mutagenesis (in which one amino acid residue is substituted for another) and deletional mutagenesis (in which sections of the polypeptide chain of a protein are eliminated) have allowed enzymologists to pinpoint the chemical groups that participate in ligand binding and in specific chemical steps during enzyme catalysis.

The study of enzymes remains of great importance to the scientific community and to society in general. We continue to utilize enzymes in many industrial applications. Moreover enzymes are still in use in their traditional roles in food and beverage manufacturing. In modern times, the role of enzymes in consumer products and in chemical manufacturing has expanded greatly. Enzymes are used today in such varied applications as stereospecific chemical synthesis, laundry detergents, and cleaning kits for contact lenses.

Perhaps one of the most exciting fields of modern enzymology is the application of enzyme inhibitors as drugs in human and veterinary medicine. Many of the drugs that are commonly used today function by inhibiting specific enzymes that are associated with the disease process. Aspirin, for example, one of the most widely used drugs in the world, elicits its anti-inflammatory efficacy by acting as an inhibitor of the enzyme prostaglandin synthase. As illustrated in Table 1.1, enzymes take part in a wide range of human pathophysiologies, and many specific enzyme inhibitors have been developed to combat their activities, thus acting as therapeutic agents. Several of the inhibitors listed in Table 1.1 are the result of the combined use of biophysical methods for assessing enzyme structure and classical pharmacology in what is commonly referred to as rational or structure-based drug design.

This approach uses the structural information obtained from x-ray crystallography or NMR spectroscopy to determine the topology of the enzyme active site. Next, model building is performed to design molecules that would fit well into this active site pocket. These molecules are then synthesized and tested as inhibitors. Several iterations of this procedure often lead to extremely potent inhibitors of the target enzyme.

The list in Table 1.1 will continue to grow as our understanding of disease state physiology increases. There remain thousands of enzymes involved in human physiology that have yet to be isolated or characterized. As more and more disease-related enzymes are discovered and characterized, new inhibitors will need to be designed to arrest the actions of these catalysts, in the continuing effort to fulfill unmet human medical needs.

Table 1.1 Examples of enzyme inhibitors as potential drugs

Inhibitor/Drug	Disease/Condition	Enzyme Target
Acetazolamide	Glaucoma	Carbonic anhydrase
Acyclovir	Herpes	Viral DNA polymerase
Allopurinol	Gout	Xanthine oxidase
Argatroban	Coagulation	Thrombin
Aspirin, ibuprofen, DuP697	Inflammation, pain, fever	Prostaglandin synthase
β-Lactam antibiotics	Bacterial infections	D-Ala-D-Ala transpeptidase
Brequinar	Organ transplantation	Dihydroorotate dehydrogenase
Candoxatril	Hypertension, congestive heart failure	Atriopeptidase
Captopril	Hypertension	Angiotensin-converting enzyme
Clavulanate	Bacterial resistance	β-Lactamase
Cyclosporin	Organ transplantation	Cyclophilin/calcineurin
DuP450	AIDS	HIV protease
Enoximone	Congestive heart failure ischemia	cAMP phosphodiesterase
Finazteride	Benign prostate hyperplasia	Testosterone-5-α-reductase
FK-506	Organ transplantation, autoimmune disease	FK-506 binding protein
Fluorouracyl	Cancer	Thymidilate synthase
3-Fluorovinylglycine	Bacterial infection	Alanine racemase
(2-Furyl)-acryloyl-Gly-Phe-Phe	Lung elastin degradation in cystic fibrosis	*Pseudomonas* elastase
ICI-200,808	Emphysema	Neutrophil elastase
Lovastatin	High cholesterol	HMG CoA reductase
Ly-256548	Inflammation	Phospholipase A_2
Methotrexate	Cancer	Dihydrofolate reductase
Nitecapone	Parkinson's disease	Catechol-*O*-methyltransferase
Norfloxacin	Urinary tract infections	DNA gyrase
Omeprazole	Peptic ulcers	H^+, K^+-ATPase
PALA	Cancer	Aspartate transcarbamoylase
PD-116124	Metabolism of antineoplastic drugs	Purine nucleoside phosphorylase
Phenelzine	Depression	Brain monoamine oxidase
Ro 42-5892	Hypertension	Renin
Sorbinil	Diabetic retinopathy	Aldose reductase
SQ-29072	Hypertension, congestive heart failure, analgesia	Enkephalinase
Sulfamethoxazole	Bacterial infection, malaria	Dihydropteorate synthase
Testolactone	Hormone-dependent tumors	Aromatase
Threo-5-fluoro-L-dihydroorotate	Cancer	Dihydroorotase
Trimethoprim	Bacterial infection	Dihydrofolate reductase
WIN 51711	Common cold	Rhinovirus coat protein
Zidovudine	AIDS	HIV reverse transcriptase
Zileuton	Allergy	5-Lipoxygenase

Chapter 2

TECHNOLOGY AND APPLICATIONS

Much before enzymes were identified as discrete biochemical entities, they found favor through their useful properties. Early applications included use of enzyme preparations in meat tenderizing and starch hydrolysis. From the very beginning, commercial enzyme applications have largely belonged to a group of hydrolytic reactions. But a few oxidative enzymes were also exploited. While this trend holds even today, examples of designer enzymes and catalysts for more complex chemical processes are being developed. The first application of diastase (α-amylase) was by Jokichi Takamine. His 1894 patent (US Patent No. 525823) describes a process to make Taka-diastase from Aspergillus oryzae. This α-amylase was useful as a digestive aid, in eliminating starchy material from textiles and laundry. In a short but succinct paper, E.F. Leuchs (in 1931) described "the action of saliva on starch." The possible practical utility of such activity was clearly anticipated by him. The last line of his report reads "it will be possible to use saliva and gastric juice of killed animals very successfully in cases of defective digestion."

The quantity and quality of an enzyme are two critical parameters that define their application and extent of use. Industrial scale processes require enzymes (often in crude form) in tons, whereas precise clinical use mandates extreme purity and minimal or no contaminating factors. Accordingly, the enzyme production costs for different end objectives vary – they can be of high volume and low cost or low volume but of high cost. For instance, medically valuable products like streptokinase and asparaginase need to be very pure and are therefore expensive. Enzyme catalysts of practical import are sought by industries in many different ways. Significant among these are screening for useful activities from the naturally abundant diversity, modifying already available enzyme properties to suit our requirements, and genetically engineering desirable properties into these catalysts. We will briefly touch upon the applications of enzymes and industrial strategies with suitable examples in this chapter. Applications of enzymes and enzyme technology can occupy volumes, and many authoritative books are available for the interested reader.

2.1 EXPLOITING NATURAL DIVERSITY

The rich biodiversity on earth goes hand in hand with naturally vast array of catalytic activities. A cleverly designed screen almost always leads to an enzyme with desired properties. Thermostable protease (from Bacillus strains) and DNA polymerase (from Thermus aquaticus) are two examples of enzymes chosen for high temperature stability. The range of natural diversity is obvious from the number of enzymes that have found niche applications in the processing of carbohydrate polymers, proteins, and lipids.

Enzymes for Bioprocessing Polysaccharides are the major biomolecules that comprise biomass on this planet. They serve two important functions – energy storage (such as starch) and structural rigidity (such as cellulose). It is therefore not surprising that enzyme technology took its roots through processes to hydrolyze these sugar polymers. Microbes (bacteria and fungi) constitute an abundant source of amylases and cellulases. Controlled hydrolysis of starch to sweeteners (and sugar substitutes) is a well-developed industry (Fig. 2.1). Various enzymes used in the starch saccharification process are α-amylases, β-amylases, glucoamylases, pullulanases, and glucose isomerase. Despite certain limitations, conversion of glucose to fructose through glucose isomerase is central to many sucrose substitutes – with distinct economic and manufacturing advantages.

Although there is an abundance of cellulose in nature, transforming cellulosic biomass into sugar has been a challenge. Concerted action of a bunch of enzymes (that constitute the "cellulase complex") is required for this . Significant advances in enzymatic processes to breakdown cellulose into fermentable sugars are being made. In the meanwhile, individual components (Table 2.1) of the cellulase complex have found application in textile and paper industry.

Proteases and lipases are next in order of significance in enzyme industry. Apart from the historical significance of papain and digestive enzymes (like trypsin and chymotrypsin), this class of enzymes has found wide-ranging applications in foods, detergents, and tanning of leather. Bacteria and fungi are ideal sources for the largescale production of proteases . Most important alkaline protease producers are Bacillus strains and fungi belonging to genus Aspergillus. Subtilisin is the best known bacterial protease additive of modern detergents. It has been extensively selected/ modified for features like pH optimum and temperature stability.

Chymosin (also known as rennin) is a milk coagulating enzyme from calf stomach, which is used for generations in cheese making. An equivalent enzyme from a microbial source was sought, and several Mucor strains are chosen to produce rennin substitutes.

Many other enzymes including lipases and pectinases are also available in industrial scale. A representative list of enzymes commonly used in industry is given in the table (Table 2.2).

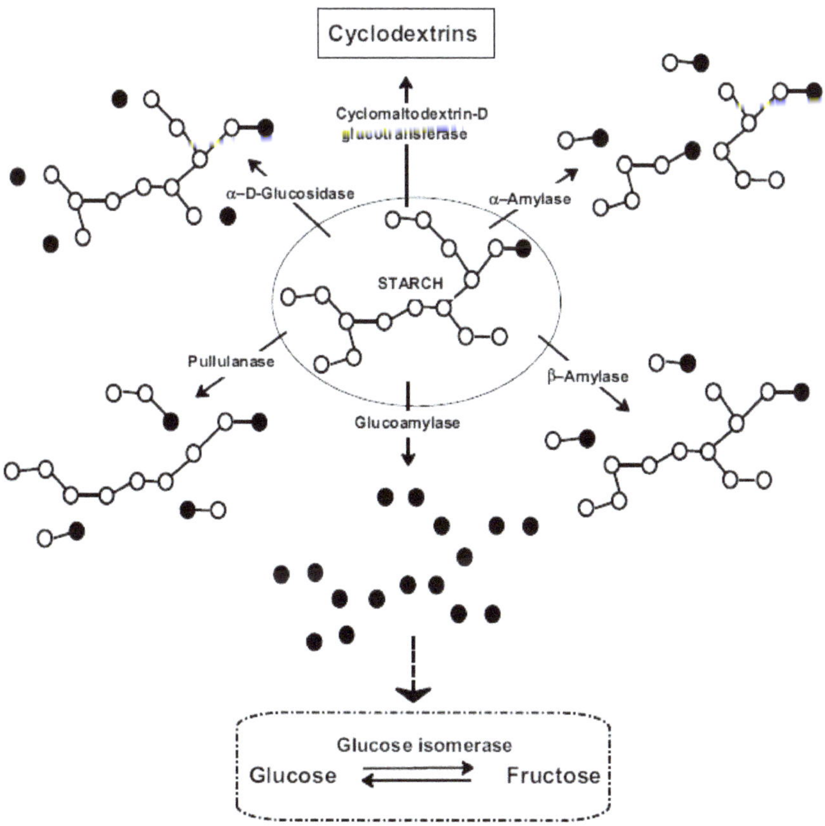

Fig. 2.1 Significant steps and enzymes employed in starch processing. Glucose residues of starch are schematically represented as circles. Filled circles indicate glucose residues whose C1-OH has not entered into a glycosidic linkage (free reducing ends). Besides glucose isomerase, all the possible enzymatic modes of dismantling starch are shown. Some combinations of these enzymes are commercially available as industrial formulation

Enzymes in Pharma and Medical Applications Pharmaceutical industry is another big beneficiary of applied enzymology. Enzymes isolated from natural sources as well as those cloned and expressed (through genetic engineering) are in use. Enzymes and their critical study serve multiple purposes in drug discovery and development.

Table 2.1 Component activities of cellulase complex and their applications

Cellulase component	Substrate specificity	Application
β-Glucosidase (Cellobiase)	Cellobiose→Glucose	Saccharification
Cellobiohydrolase I (CBH1)	Cellulose→ Cellobiose (exo – Nonreducing end)	Biomass conversion
Cellobiohydrolase II (CBH2)	Cellulose→ Cellobiose(both exo and endo)	Biomass conversion
Endoglucanase I (EG1)	Cellulose (endo)	Textile/fabric softening,
		Biopolishing
Endoglucanase II (EG2)	Cellulose (endo)	Textile/fabric softening,
		Biopolishing
Xylanase	Xylan	Paper pulp deinking
All components	Cellulose and Xylan	Feed/fodder, biomass
		Conversion

Table 2.2 Large-scale use of enzymes in industry

Enzyme	Application
Acting on carbohydrates	
Amylases	Starch processing
Cellulase complex	Biomass conversion, textile industry
Pectinases, esterases	Food industry, fruit juice, brewing
Glucose isomerase, invertase	High fructose syrups, invert sugar
Acting on proteins	
Papain, pepsin	Meat and leather processing, treating dough
Rennin, chymosin	Cheese making
Subtilisin	Detergents, leather and wool processing
Acting on lipids and esters	
Lipases	Food and detergent industry, cocoa butter
Acting on antibiotics	
Penicillin acylase	Produce 6-aminopenicillanic acid (6-APA)

The concept of screening for enzyme inhibitors was first adopted by Hamao Umezawa's group in Japa. Since then, many enzyme inhibitors have been discovered (few are listed in Table 2.3) and are in use. For instance, preventing absorption of dietary fat (triglycerides) can be a possible strategy to control obesity. An appropriate lipase from the digestive juices could serve as a target for this screening (Fig. .2).

Often the active chemical entity obtained from an enzyme screen may not find direct application. These lead compounds (inhibitors) are suitably altered/derivatized to achieve better bioavailability and reduce toxicity. An in-depth kinetic analysis of enzyme inhibition is at the heart of modern drug discovery programs.

Table 2.3 Examples of enzyme-targeted screens for active principles

Enzyme target	Screening outcome	End use
Pepsin	Pepstatin	Ulcers
Angiotensin converting enzyme	Captopril	Hypertension
HMG CoA reductase	Lovastatin	Hypercholesteremia
α-Amylase	Acarbose	Diabetes
Triacylglycerol lipase	Orlistat (lipostatin)	Obesity
Acetylcholine esterase	Rivastigmine	Alzheimer's disease
β-Lactamase	Clavulanic acid	Combination therapy

Digestion of FAT

↓

Lipase

(involved in hydrolysis of triacylglycerols)

↓

Inhibition

(lipase activity - enzyme assay)

↓

Inhibitor Screen

(natural or synthetic chemical libraries)

↓

Lipostatin

(active principle)

Fig. 2.2 Flow chart outlining the design of a lipase inhibitor screen

☆ Besides being targets for inhibitor screens, many enzymes are employed as catalysts for synthesis. The whole range of β-lactam antibiotics available today includes a large number of semisynthetic penicillins and cephalosporins. While penicillin G is produced by fermentation, 6-aminopenicillanic acid – an important precursor for semisynthetic penicillins – is derived from it (Fig. 2.3). Penicillin acylase is a valuable commodity in the large-scale production of 6-aminopenicillanic acid.

☆ Because of their catalytic potential coupled with specificity, many enzymes are used as exquisite analytical tools. Alkaline phosphatase and peroxidase are two reporter enzymes of extensive history in ELISA (enzyme-linked immunosorbent assay). Here the desired specificity of interaction (through antibodies) is coupled to the signal amplification provided by enzyme catalysis. Enzymes as antibodyconjugates find routine use in detection of DNA/RNA/protein on blots. Taq DNA polymerase is extensively employed for DNA amplification through polymerase chain reaction (PCR). A number of metabolites are analyzed (in a clinical setting) are not acted upon by the β-lactamase. Penicillin acylase is used to produce 6-aminopenicillanic acid (6-APA). 6-APA is an important precursor to make semisynthetic penicillins, both through enzymatic and chemical routes through assays involving enzymes. A list of more commonly used enzymes and the corresponding analytes are given in Table 2.4. Some of these enzymes are also employed as a component of biosensors (see below).

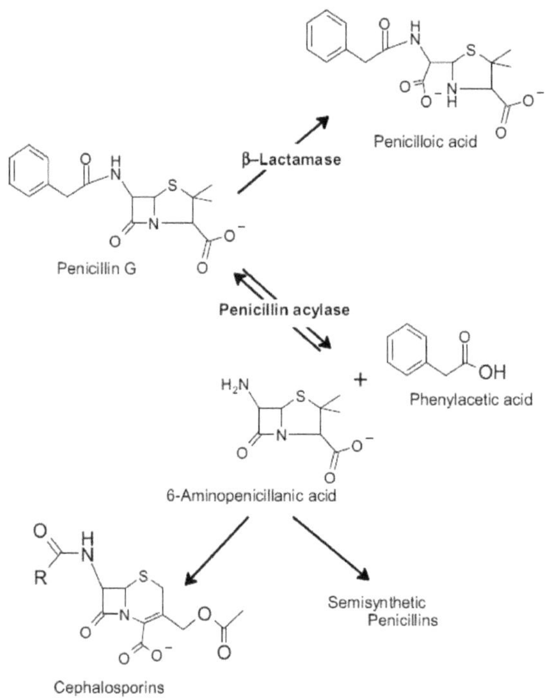

Fig. 2.3 Enzymes and steps relevant to penicillin (β-lactams) industry. Antibiotic resistance is often due to a β-lactamase; better antibiotics may be evolved by screening for novel structures that

☆ Enzymes have found medical applications in terms of diagnosis as well as therapy. A few enzymes find direct application as therapeutic agents in medicine. Best known examples from the market include diastase (α-amylase, digestive aid), asparaginase (leukemia, antitumor therapy), rhodanese (cyanide poisoning), and streptokinase (medication to dissolve blood clots). However, a large number of enzymes are routinely monitored as clinical markers (Table 2.5). Enzyme profiles from serum, amniotic fluid, urine, etc. are monitored, and their levels are often correlated with disease conditions and aid in the diagnosis of disease.

Table 2.4 Examples of enzymes for metabolite analysis

Enzyme	Analyte detected/estimated
Catalase	Hydrogen peroxide
Glucose oxidase	Glucose
Hexokinase	Glucose
Alcohol dehydrogenase	Ethanol
Lactate dehydrogenase	Lactate/pyruvate
Luciferase	ATP
Urease	Urea
Cholesterol oxidase	Cholesterol

Table 2.5 Examples of enzymes as clinical markers

Enzyme	Used as marker for
Lactate dehydrogenase (H4 isoform)	Heart diseases
Glutamate-oxoglutarate transaminase (SGOT), glutamate-pyruvate transaminase (SGPT)	Liver function
Creatine kinase	Myocardial infarction, skeletal muscle damage
Lactase	Lactose intolerance
Hexosaminidase A	Tay-Sachs disease
Acid phosphatase	Prostate cancer
Phenylalanine hydroxylase	Phenylketonuria (PKU)

☆ **Enzymes and Issues of Safety** Industrial preparation and use of enzymes come with its own safety and regulatory issues. Potential hazards from exposure to large quantities of a given enzyme include allergenicity, functional toxicity, chemical toxicity, and source-related contaminants. Large-scale manufacture of enzymes often ends with partially pure yet enriched preparations. Such material could contain potentially toxic chemicals (like mycotoxins) carried over from the source. Not all microorganisms are safe, and their trace contamination in the final enzyme preparations requires attention. For enzymes to be used in food ingredients, they must be GRAS (generally recognized as safe). Some enzymes like proteases are potentially dangerous – particularly upon exposure of sensitive tissues to concentrated preparations. Since enzymes are proteins, they can be potent allergens. Repeated exposure through inhalation or skin contact can trigger severe allergic response. Enzyme preparations especially handled in the form of dust, dry powder, or aerosol are harmful and must be avoided. Many issues of safety regarding free enzyme preparations may be overcome by using them in the immobilized, granulated, or encapsulated form.

2.2 MODIFYING ENZYMES TO SUIT REQUIREMENTS

Despite the vast natural diversity of biological catalysts, significant technology has developed to alter the properties of available enzymes. This tinkering has involved facets of their immobilization, chemical modification, genetic engineering, or their use in nonaqueous solvents.

Immobilization for Better Use Most natural enzymes isolated are in water soluble state. They cannot be stored in this form for long, often due to instability. Their immobilization is one way to enhance their shelf life . In addition, immobilized enzymes are easy to recover and amenable to repeated use. This is an important consideration when the cost of enzyme is very high. The characteristics of the matrix are very critical in determining the performance of the immobilized enzyme system. These supports may be inorganic or organic according to the nature of their chemical composition. The physical characteristics like mean particle size, swelling behavior, mechanical strength, etc. decide the technical conditions in which the system is used. Enzymes may be quarantined on the matrix either irreversibly (by covalent bonding, entrapment, microencapsulation, cross-linking, etc.) or reversibly (by adsorption, ionic binding, affinity binding, disulfide bonds, or chelate/ metal binding). The cost associated with the process of immobilization determines whether it is economically viable to do so.

Different means are adopted for enzyme immobilization in practice, and only a few are represented in Fig. 2.4. The field of enzyme immobilization technology and its applications has grown vastly over the years. Many books and volumes (Methods in Enzymology series) are available for detailed reference . This section covers very briefly on this applied aspect of enzymes, and the reader is encouraged to refer the more specialized literature for the purpose.

The choice of immobilization method depends on the type of the enzyme and the nature of applications in question. Non-covalent confinement (like physical entrapment, microencapsulation, or electrostatic adsorption) methods at times may lead to enzyme leaching during operation. Covalent anchoring of enzymes on the other hand requires bifunctional cross-linking reagents and suitable functional groups on the enzyme surface.

These functional groups must not be critical for enzyme activity however. A great deal of sophisticated chemistry has been developed to activate inert organic/inorganic polymers for subsequent enzyme immobilization. Carrier-bound (covalently linked to polymers; figure above) penicillin acylase is highly effective in the preparation of 6-APA and permits economic recycling of the catalyst. Penicillin acylase immobilized cassettes are available that function at >99% conversion efficiency even after 1500 cycles of use. Glutaraldehyde is used to cross-link glucose isomerase – the cross-linked material can be reused many times in the commercial process for production of fructose syrups from glucose. Detergent protease components act as allergens, and this problem was overcome through microencapsulation; the dustless protease preparations have reduced this risk. Some of the enzymes successfully immobilized for application are shown in Table 2.6.

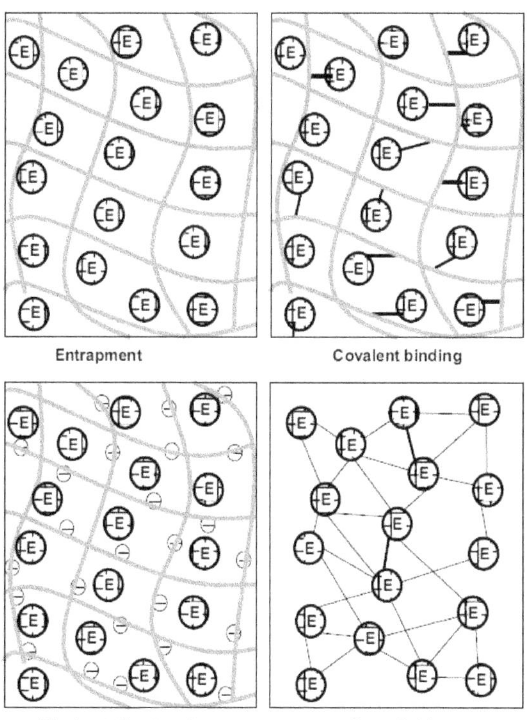

Fig. 2.4 Four different modes of enzyme immobilization

Table 2.6 Industrial uses of immobilized enzymes

Enzyme	Product
Aspartate ammonia lyase	L-Aspartic acid
Hydantoinase	L- and D-Amino acids
Thermolysin	Aspartame
Lactase	Lactose-free milk and whey
Invertase	Invert sugar
Glucose isomerase	High-fructose syrup
Lipase	Cocoa butter substitute
Penicillin acylase	6-APA and penicillins
Nitrile hydratase	Acrylamide

It is expected that the substrate specificity and the catalytic potential of the enzyme should not be unfavorably affected upon immobilization. The changed microenvironment could affect the enzyme stability to heat, pH, and proteolytic enzymes. A reduction in its activity may also occur due to associated conformational changes in the enzyme. The pH optimum of an immobilized enzyme may change by as much as 2.0 pH units due to microenvironment effects. It is observed that an anionic carrier (matrix) pushes the pH optimum to a more alkaline value, while a cationic carrier moves it to a more acidic value. These matrix (electrostatic field) effects are due to changes in the degree of ionization of amino acid residues on the enzyme; this effect is usually not observed in medium with high ionic strength – as salt ions counter the charges on the carrier.

The kinetic behavior of an immobilized enzyme can differ significantly from the free enzyme in solution. Both maximal velocity (Vmax) and Michaelis constant (KM – apparent affinity for the substrate; of the enzyme may be affected. The apparent KM significantly decreases when the carrier used is of opposite charge to that of the substrate – largely due to electrostatic interactions causing the substrate to be at a higher concentration near the carrier. In general, partition effects (arising from ionic, electrostatic, or hydrophobic interactions) cause changes in the local concentration of the ligands/molecules which in turn affect the kinetic constants. The apparent KM of an immobilized enzyme is also affected by diffusion factors. Both the diffusion of the substrate to and the diffusion of the product away from the enzyme influence enzyme activity. Restricted diffusion of bulk substrate to the immobilized enzyme leads to significantly higher KM value. Poor product diffusion, away from the enzyme, obviously results in an inhibited enzyme. If diffusion constrains the highest substrate concentration achievable, then the immobilized enzyme may display an apparent Vmax lower than the free soluble enzyme.

The objective of using enzyme-catalyzed reactions in industry is to convert given amount of the substrate maximally into the product. Integrated form of the rate equation (Michaelis-Menten equation) is better suited to analyze such systems.

However, the integrated form of the rate equation is also valid only if (a) substrate concentrations far exceed that of the enzyme, (b) the overall reaction is irreversible, (c) enzyme is stable over the time period, (d) no product or substrate inhibition occurs, and (e) the system remains properly mixed . Some of these factors can be tackled through choice of enzyme reactor configuration and process design. Continuous industrial processes could employ immobilized enzyme in stirred-tank, packed-bed, or fluidized-bed reactors or the enzyme immobilized on membranes, hollow fibers, or tubes. The flow of the reaction mixture may be exploited to input fresh substrate, remove the product formed, adjust pH, etc. Membrane cassettes of penicillin acylase have made the production of 6-APA very efficient – the reaction mixture pH can be maintained by dosing NaOH (pH stat) and continuously removing the product formed.

Enzyme Biosensors Biosensors are used to determine/monitor the concentration of substances (often metabolites) of biological interest. Enzymes play a major role in such analytical devices – to convert a biological response into an electrical signal. Specificity and signal amplification (catalysis) of an immobilized enzyme are ideally suited for this purpose. Besides the fact that they may be reused, the immobilized enzymes (with an elevated KM value) give proportional change in reaction rate over a substantial linear range of the substrate concentrations. Also, often, these rates are independent of pH, temperature, ionic strength, and inhibitors – features advantageous in metabolite measurements in real analytical samples.

The critical component of the biosensor is the transducer which converts the outcome of an enzyme reaction into a measurable signal. Biosensors may have a transducer to exploit enzyme reactions with (1) heat generation (calorimetric biosensor, to measure H2O2 with catalase), (2) release or absorption of ions (potentiometric biosensor, glucose with glucose oxidase and urea with urease), (3) production of a current (amperometric biosensor – glucose with glucose oxidase, alcohol with alcohol oxidase, and cholesterol with cholesterol oxidase), and (4) absorption or emission of light (optical biosensor, peroxides with horseradish peroxidase and ATP with luciferase). Paper enzyme strips are also in use to measure/detect substances through colorimetry. Enzyme biosensors today occupy a substantial analytical market in health care, food industry, and environmental monitoring.

Function in Organic Solvents – Nonaqueous Enzymology Living systems are intimately linked with their aqueous environment. Being biological catalysts, enzymes are exquisite products of long biological evolution. Enzymes are easily denatured by organic solvents. However, some enzymes can tolerate high concentrations of water-miscible organic solvents in their aqueous surroundings.

Even in a water-immiscible nonpolar solvent, enzymes do need a monolayer of water molecules covering their exposed surfaces and the active site . In this arrangement – with an essential water layer – they can continue to function as catalysts even in an organic solvent. The early work of Bourquelot and others (since 1913) showed that few enzymes could act in the presence of >80% of organic solvents such as ethanol or acetone. This example may very well be the forerunner of nonaqueous enzymology and use of enzymes in organic solvents for synthetic applications. The synthesis of a glucoside highlighted very early the synthetic capabilities of an enzyme.

Enormous interest in biocatalysis in nonaqueous phase was triggered due to the merits of good enantioselectivity, reverse thermodynamic equilibrium, and no waterdependent side reactions. For most organic reactions, water is not an ideal solvent.

But enzymes have evolved for the catalysis of reactions in water. It would be very useful to have enzymes perform catalysis in nonaqueous media. There are added advantages if this is made possible:

- ☆ Many substrates are more soluble in organic solvents than in water. Large initial concentrations can be achieved with hydrophobic substrates that are sparingly soluble in water. They continuously diffuse into the active site from the bulk solution (which is nonaqueous). Some products are labile in aqueous media, and thus water-dependent side reactions can be minimized.
- ☆ Enzymes naturally find themselves in an aqueous environment with 55.5 M of water. Introducing them into nonaqueous environment means a drastically reduced water concentration! There are important consequences of lower water activity – particularly with reactions in which water is added or removed. For instance, peptide bond synthesis involves elimination of water – a thermodynamically uphill task in water. However, this step is favored in a nonaqueous environment. As a consequence a hydrolytic enzyme could be endowed with synthetic potential. Most hydrolytic reactions – generally irreversible in water – can be made reversible. Esterases, glycosidases, and proteases can be used for synthesis.

All water-soluble enzymes possess a small but significant amount of strongly bound water – thereby resulting in a two-phase system within the nonaqueous bulk medium. The water activity – aw – is defined as the partial vapor pressure of water in a substance divided by the standard state partial vapor pressure of water. It is indicative of the extent of water content around the enzyme molecules. Enzyme activity in nonaqueous media depends on the magnitude of aw because it affects the extent of this bound water. Lower aw may lead to a rigid enzyme with limited thermal motion and associated thermal stability. The miniscule water pool around the enzyme in an organic solvent (between 50 and 500 water molecules per enzyme molecule) retains the pH of the last aqueous solution from where it is derived. Therefore, it appears as if the enzyme remembers and functions in that pH (pH memory).

The enzyme is inactivated if the tightly bound water layer is stripped off or diluted by the organic solvent phase. The stability or inactivation of an enzyme is thus dictated by the polarity of the solvent used. A useful measure of this polarity is log P – the logarithm of the partition coefficient of the organic solvent (X) between n-octanol and water. Log P, the partition coefficient, is a measure of hydrophobicity of organic solvent. The lower the P value, the more polar (hydrophilic) is the solvent.

$$\log P = \log\left(\frac{[X_{Octanol}]}{[X_{Water}]}\right)$$

As an empirical rule, enzymes are generally inactivated by solvents with log $P < 2$ but are little affected by more hydrophobic solvents with log $P > 4$.

Industrial focus on the use of enzymes in nonaqueous environments and reverse micelles systems is on the rise, with applications in foods, medicine, and industry. Being chiral catalysts, they are often valuable in resolving racemic mixtures of important drugs and intermediates in pharmaceutical industry. In practice it is common to find applications of nonaqueous enzymology in transglycosylation, transesterification, and transpeptidation reactions. An important example of the reversal of peptide bond hydrolysis is aspartame (α-L-aspartyl-L-phenylalanyl-Omethyl ester) synthesis. The protease thermolysin is used to condense L-aspartic acid with the methyl ester of L-phenylalanine and produce aspartame (sweetener and a sugar substitute). Similarly, suitable glycosidases are used to synthesize cyclodextrins. Interesterification reactions with lipases in nonaqueous solvents have found many applications. Consider an esterase catalyzing the hydrolytic reaction:

$$R - COOX + H - OH \rightarrow R - COOH + X - OH$$

The same enzyme, in the presence of another alcohol (Y-OH), but in the absence of water (low aw), can bring about efficient catalysis of an ester exchange shown below:

$$R - COOX + Y - OH \rightleftarrows R - COOY + X - OH$$

For example, isoamyl acetate (a banana fruit flavor) is produced from ethyl acetate and isoamyl alcohol. Yet another interesting patented example of transesterification with a lipase is the preparation of cocoa butter substitute from palm oil.

2.3 GENETIC ENGINEERING AND ENZYMES

Large-scale production of enzymes is often a prerequisite for most applications.

Obtaining them from animal and plant material - though of historical importance - has become progressively difficult for economic and ethical reasons. Therefore, the extant microbial biodiversity is routinely screened for enzymes with similar/ desirable properties. Microbial rennin (a substitute for chymosin) produced from Mucor spp. is a case in point. Another exciting option is to produce the required enzyme through recombinant DNA technology, preferably in a microbial host.

The formidable tools of genetic engineering have allowed the expression and management of enzyme structures almost at will. Detailed recipes of recombinant DNA techniques are available in many texts and protocol books, while we sketch a brief outline of these steps here. Systematic manipulation of the DNA sequence at the molecular level is the essence of genetic engineering. This means we can cut and patch DNA fragments/ gene(s) for any enzyme protein from diverse living organisms. These recombinant DNA molecules - capable of expressing a natural or a mutant enzyme protein - can then be moved into a suitable host for protein expression. When this genetic information is expressed (by transcription and translation) in the new host, it can produce an enzyme protein that is even foreign to it! A general strategy for genetic engineering is outlined in Fig. 2.5. There are essentially four stages to this powerful technique.

Isolation of the Gene/ORF/cDNA for the Enzyme Protein Restriction endonucleases of different sequence specificities are employed to cut out the desired gene or the ORF (open reading frame) encoding that enzyme from a given source of DNA. The required DNA fragment may be amplified by PCR (polymerase chain reaction) or the corresponding cDNA may be obtained through reverse transcriptase-PCR of the relevant mRNA. Alternatively, the required DNA may also be synthesized chemically - if the desired nucleotide or amino acid sequence is known.

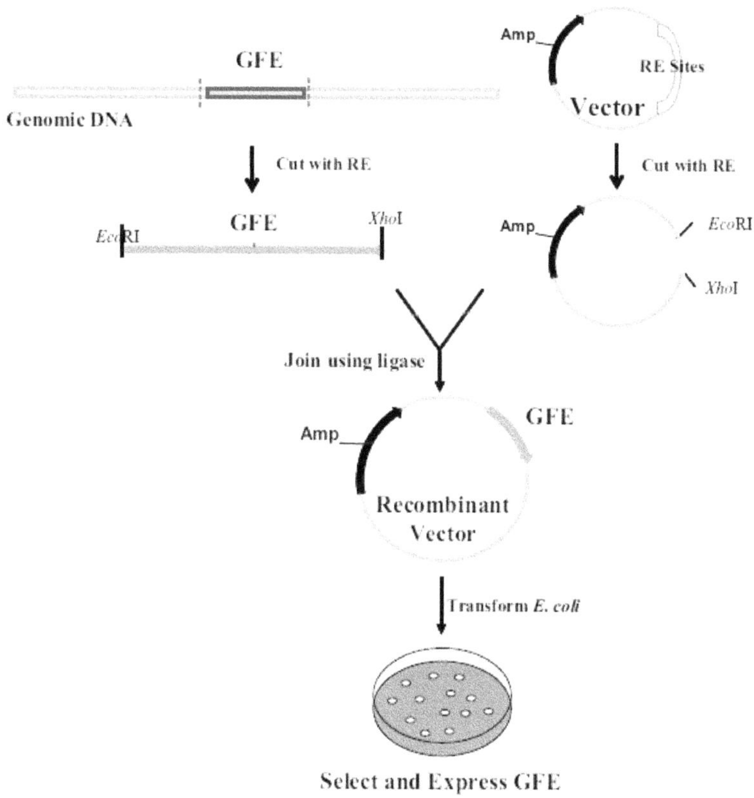

Fig. 2.5 A general genetic engineering strategy for enzyme expression

Insertion of the Gene Into an Expression Vector The gene/ORF/cDNA/ is integrated (linked) into another piece of DNA, the vector DNA, in order to promote its uptake and replication in a suitable host organism. In general, the DNA vectors are designed (engineered) to have a unique site for restriction endonuclease and carry a marker to facilitate selection of genetically modified host cells (transformants). The chimerical recombinant DNA is introduced in appropriate host for expressing the enzyme protein. Typically, a bacterial plasmid vector is used to transform bacteria (Fig. 2.5). For animal and plant cells, their respective viral DNAs are often used as vectors.

Transformation of the Host Cell The recombinant DNA vector is introduced into a host cell either directly (by the process of transformation) or by infecting it using a viral vector.

Detection of the Inserted Gene The presence of the foreign DNA insert in the host is detected by molecular tools like Southern blotting (DNA-DNA hybridization), PCR, etc. The functional expression of the recombinant DNA may be directly monitored through enzyme activity and the presence of the protein.

At its simplest level, genetic engineering tools allow us to produce any enzyme protein in the common bacterium Escherichia coli. One could choose from bacteria, yeast, plant, or mammalian cells as hosts for optimal expression and/or appropriate posttranslational processing (like glycosylation, etc.) of the protein. Expression in a homologous host is generally successful. One of the principal reasons that an enzymologist manipulates DNA blueprint of an enzyme is to modulate the existing feature or to create new ones. Approaches like site-directed mutagenesis (SDM) and directed evolution of enzymes are routine nowadays.. In addition, genetic engineering has made optimal enzyme production possible in many different ways. We will simply illustrate the field with a few examples:

1. Enzyme overproducing strains have been constructed by suitably overcoming regulation at the level of feedback inhibition, transcription, translation, or secretion of the enzyme protein. Expression of amylases is often under the control of carbon catabolite repression (glucose repression). Yet another bottleneck with respect to protease production is nitrogen metabolite regulation. Both bacterial and fungal strains, with mutations for deregulation, constitutive expression, and hypersecretion, find utility in enzyme production. Suitable genetically stable mutant strains combining many such features have found their place in enzyme industry. Individual component activities of the cellulase complex and their specific combinations find industrial applications (Table 2.1). Producer strains (like Trichoderma reesei) overexpressing individual activities as well as deleted in each one of them are available. Well-defined cocktail of cellulase components are suitable in textile industry and in biomass conversion.
2. Heterologous expression, even of mammalian or plant enzymes, in a convenient microbial host. The recombinant chymosin (bovine) was produced in a fungus by introducing a stable expression construct. The difficult challenges however include ability to obtain economically viable levels of secretion, stability of the expressed protein, and attaining proper posttranslational modifications like glycosylation, if any.
3. Once the structural gene for the enzyme is cloned and expressed, it is feasible to generate mutant forms of this enzyme. Tinkering with enzyme properties like stability, pH optimum, specificity, and regulatory features is possible. Protein and enzyme engineering through site-directed mutagenesis has found direct applications in the enzyme industry. Subtilisin was engineered for better stability.

Chapter 3

ENZYME NOMENCLATURE AND CLASSIFICATION

3.1 WHAT IS IN THE NAME?

All enzymes are proteins but not all proteins are enzymes. Catalysis by RNA molecules (the so-called ribozyme) has expanded the realm of biological catalysts to beyond proteins. While few RNA catalysts have been recognized, the vast majority of enzymes that we come across in biology are proteins. Evolution has selected L-isomers of 20 amino acids to build proteins. This has put limits on the number and nature of available reactive chemical groups/functions that could be recruited for catalysis. Proteins are rich in nucleophilic groups, but electrophiles are poorly represented. In some instances, the protein component alone is inadequate to catalyze a given reaction. Nature therefore recruited many nonprotein components called cofactors to generate a functional catalyst. In such enzymes, the inactive protein component without the cofactor is termed the apoenzyme and the active enzyme, including the cofactor, the holoenzyme. The cofactors may be either metal ions (e.g., Mn[II] in arginase, Ni[II] in urease, and Ca[II] in DNase I) or coenzymes (organic molecules like pyridine nucleotides NAD+ and NADP+, flavin adenine dinucleotide (FAD), pyridoxal phosphate (PLP), thiamine pyrophosphate (TPP), biotin, cobamide, and heme). Binding of a cofactor to its cognate apoenzyme could exhibit a range of binding strength. A very tightly bound cofactor – which is difficult to remove without damaging the enzyme – is also known as a prosthetic group. Quite often prosthetic groups are covalently bound to the apoenzyme – lipoamide of transacylase is an example.

3.2 ENZYME DIVERSITY AND NEED FOR SYSTEMATICS

The middle of the twentieth century saw an exponential increase in research on enzymes. Sooner than later the number of new enzymes reported crossed manageable limits for an individual. As a consequence, in some cases, the same (or similar) enzyme activities were given different names. Some names like catalase give very little indication of the nature of the reaction they catalyze. Systematic classification, cataloging, and nomenclature of enzymes therefore became a necessity. This was easier recognized than done. Enzymes could be grouped according to any of the following considerations:

a. Occurrence and/or source of the enzyme: Laccase obtained from Japanese lacquer tree, papain from papaya, and horseradish peroxidase are examples of three such plant enzymes. Similarly a number of digestive enzymes are isolated from pancreatic juice such as trypsin, chymotrypsin, carboxypeptidase, lipase, etc. the common source of lysozyme is from hen egg white (HEW).
b. Nature of the substrate on which the enzyme acts: They could be classified into enzymes hydrolyzing (or acting on) proteins, carbohydrates, lipids, etc.
c. Based on cofactor requirement: Typically many enzymes are simply proteinaceous in nature. However those depending on a cofactor could be listed into separate groups like thiamine pyrophosphate (TPP) enzymes, pyridoxal phosphate (PLP) enzymes, metalloenzymes, etc.
d. Common functional context: One could, in principle, group enzymes belonging to discrete pathways like glycolytic enzymes, enzymes of histidine biosynthesis, etc. they may also be grouped as soluble, membranebound, or belonging to organelles such as mitochondria, etc.
e. Nature of the overall reaction catalyzed: An enzyme can be assigned to a group by considering the type of the reaction it catalyzes. For instance, they may catalyze oxidation, hydrolysis, etc.
f. The mechanism of reaction: The intimate mechanism of the reaction at the enzyme active site and the nature of intermediate complexes with the enzyme may be considered. For example, proteases may be classified depending upon whether an enzyme-bound acyl-enzyme intermediate is formed or not.

It should be obvious from the above list that a systematic, meaningful classification and cataloging of all the enzymes has not been easy. The problem is compounded by the enormous diversity of enzyme structures and activities. A typical RNA hydrolysis is achieved through a protein (RNase A), a protein-RNA complex (RNase P) or RNA alone (ribozyme). The peptide bond hydrolysis is possible with enzymes that are efficient in acidic pH or alkaline pH, require a divalent metal ion, contain a serine -OH or cysteine -SH, etc. Enzymatic decarboxylation of histidine may recruit pyridoxal phosphate or in a more primitive form may simply use a bound pyruvate. Pyridoxal phosphate bound to glycogen phosphorylase serves more of a structural role rather than function as a cofactor. Alkyldihydroxyacetonephosphate synthase (an enzyme involved in the biosynthesis of ether phospholipids) uses FAD for a non-redox reaction. Clearly, no single criterion listed above would be satisfactory.

3.3 ENZYME COMMISSION: RECOMMENDATIONS

Considering the diversity of enzyme sources, reactions, and mechanisms, it became apparent that a formal system of nomenclature and classification was required. "Enzyme Commission" was appointed by the International Union of Biochemistry to address this issue. Its first report, published in 1964, forms the basis of present system of classification. This system of classification is being updated periodically with updates made in 1972, 1978, 1984, and 1992. There are also many electronic supplements such as Supplement 14 of 2008 . The most recent information and guidelines on enzyme nomenclature may be found at the official web site of the International Union of Biochemistry and Molecular Biology (IUBMB) – http://www.chem.qmul.ac.uk/iubmb/enzyme/. The EC classification is universally accepted with a unique name and EC number for each enzyme. By this means names for every enzyme (listed with trivial name) could be rationalized and also given an EC catalog number .

The nature of the overall reaction catalyzed by the enzyme – expressed by the formal equation – forms the basis of EC classification. Clearly the intimate mechanism of the reaction and the formation of intermediate complexes with the enzyme, if any, are not considered. The Enzyme Commission defined six general categories of reactions, thereby assigning an enzyme to one of the six classes. A number of unique enzymes, represented in each of these six classes , are shown in Fig. 3.1. Expectedly, enzymes for the redox and hydrolytic reactions are the most represented group. These six classes are oxidoreductases, transferases, hydrolases, lyases, isomerases, and ligases. Each of these six classes is further divided into a number of subclasses and sub-subclasses, according to the nature of the reaction catalyzed. In this classification each enzyme is given a code number consisting of a four-number system. On this system the first number indicates the main class and the second and third show the subclass and sub-subclass, respectively, thus defining the type of reaction. The fourth number is the actual number of that enzyme within its sub-subclass. For example, alcohol dehydrogenase is given the code "EC 1.1.1.1." The first number indicates that it belongs to oxidoreductase class (EC 1.x.x.x). Within this class, enzymes acting on CH-OH group of donors bear the same subclass number (EC 1.1.x.x). Within this subclass, enzymes that use NAD or NADP as electron acceptor are given the number EC 1.1.1.x. Since alcohol dehydrogenase is the first enzyme in this category, it gets its fourth number (EC 1.1.1.1). Similarly, all carboxylesterases have the same first three digits in their EC code (EC 3.1.1.x). The fourth digit however distinguishes them by the actual carboxylic ester they hydrolyze.

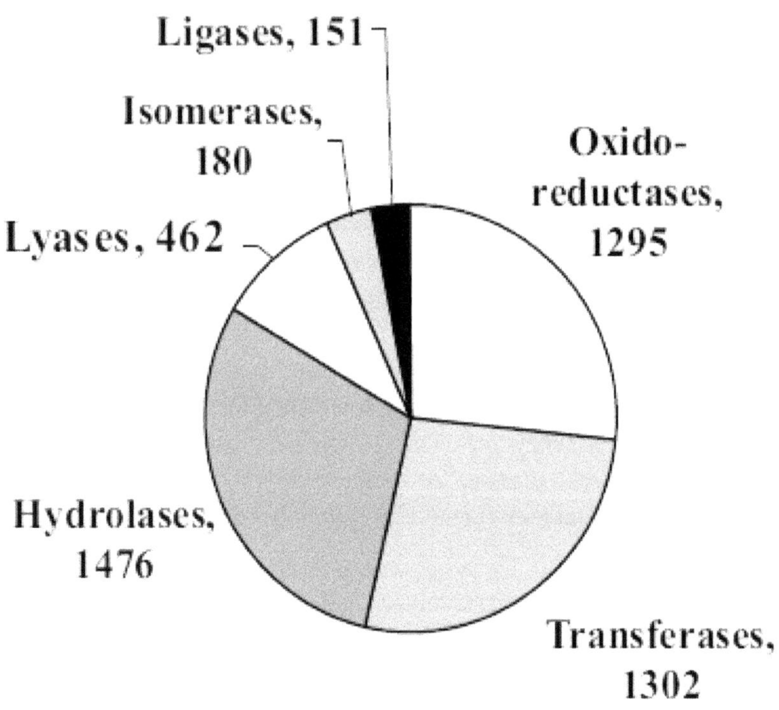

Fig. 3.1 Distribution of enzymes into six different classes according to EC classification.

Systematic name is assigned to each enzyme by the Commission, in addition to an accepted trivial name. This name includes the name or names of substrates followed by a reaction name that ends in "-ase." Because such systematic names can at times be too long and unwieldy, the Commission has also made recommendations for the use of trivial names. However, for the group (EC 3.3.3.x) of common proteases like pepsin, trypsin, papain, etc., it has not yet been possible to find acceptable systematic names. Enzyme Commission nomenclatures for enzymes representative of each class and those enzymes commonly referred to in this book are given in Table 3.1. The universally accepted EC classification and enzyme codes are finding place (and direct utility) in a number of databases describing enzymes, genes, genomes, and metabolic pathways. Some of these databases are listed in Table 3.2.

Table 3.1 Enzyme commission nomenclatures for representative/common enzymes

EC No.	Systematic name	Trivial name	Reaction
1. *Oxidoreductases: Loss or gain of electrons by substrates*			
1.1.1.1	Alcohol:NAD^+ oxidoreductase	Alcohol dehydrogenase	Alcohol+$NAD^+ \rightleftarrows$ Aldehyde+NADH+H^+
1.1.1.27	L-lactate:NAD^+ oxidoreductase	Lactate dehydrogenase	L-lactate+$NAD^+ \rightleftarrows$ Pyruvate+NADH+H^+
1.2.1.2	Formate:NAD^+ oxidoreductase	Formate dehydrogenase	Formate+$NAD^+ \rightleftarrows CO_2$ + NADH+H^+
1.4.1.4	L-Glutamate:$NADP^+$ oxidoreductase (deaminating)	NADP-glutamate dehydrogenase	L-Glutamate+H_2O + $NADP^+ \rightleftarrows$ 2-Oxoglutarate+NH_3 + NADPH +H^+
1.11.1.6	Hydrogen peroxide:Hydrogen peroxide oxidoreductase	Catalase	$H_2O_2 + H_2O_2 \rightleftarrows O_2 + 2H_2O$
2. *Transferases: Transfer of a reactive group from a donor substrate to an acceptor substrate*			
2.1.2.1	L-Serine:Tetrahydrofolate 5,10-hydroxymethyl-transferase	Serine hydroxymethyltransferase	L-Serine+Tetrahydrofolate$\rightleftarrows$ Glycine+5,10-Methylenetetrahydrofolate
2.1.3.1	Methylmalonyl-CoA: Pyruvate carboxytransferase	Methylmalonyl-CoA carboxytransferase	Methylmalonyl-CoA + Pyruvate$\rightleftarrows$ Propionyl CoA + Oxaloacetate
2.4.1.8	Maltose:Orthophosphate glucosyltransferase	Maltose phosphorylase	Maltose+Orthophosphate$\rightleftarrows$ β-D-Glucose 1-phosphate+D-Glucose
2.6.1.1	L-Aspartate:2-Oxoglutarate aminotransferase	Aspartate aminotransferase	L-Aspartate+2-Oxoglutarate$\rightleftarrows$ Oxloacetate+L-Glutamate
2.7. 1.1	ATP:D-Hexose 6-phosphotransferase	Hexokinase	ATP + D-Hexose$\rightleftarrows$ ADP + D-Hexose 6-phosphate
2.7.7.16	Ribonucleate pyrimidine-nucleotido-2′-transferase (cyclizing)	Ribonuclease	Transfers 3′-phosphate of a pyrimidine nucleotide residue of polynucleotide from 5′ position of the adjoining nucleotide to the 2′ position of the pyrimidine nucleotide itself, forms a cyclic nucleotide
3. *Hydrolases: Introduction of elements of water into a substrate*			
3.1.1.1	Carboxylic ester hydrolase	Carboxylesterase	Carboxylic ester+$H_2O\rightleftarrows$ Alcohol+Carboxylate
3.1.3.1	Orthophosphoric monoester phosphohydrolase	Alkaline phosphatase	Orthophosphoric monoester $H_2O\rightleftarrows$ Alcohol+Orthophosphate
3.2.1.1	α-1,4-Glucan 4-glucanohydrolase	α-Amylase	Hydrolyzes α-1,4-glucan links in starch
3.2.1.17	Mucopeptide N-acetylmuramylhydrolase	Lysozyme	Hydrolyzes β-1,4-glucan links in peptidoglycan
3.4.4.4	Not possible yet	Trypsin	Hydrolyzes peptides, amides and esters of aromatic L-amino acids

(continued)

EC No.	Systematic name	Trivial name	Reaction
3.5.3.1	L-Arginine amidinohydrolase	Arginase	L-Arginine+H_2O⇄ L-Ornithine+Urea
3.7.1.1	Oxaloacetate acetylhydrolase	Oxaloacetase	Oxaloacetate+H_2O⇄ Oxalate+Acetate
4. Lyases: Elimination of a group with double bond formation or addition of group to double bond			
4.1.1.1	Pyruvate carboxy-lyase	Pyruvate decarboxylase	Pyruvate⇄ Acetaldehyde+CO_2
4.2.1.1	Carbonate hydro-lyase	Carbonic anhydrase	H_2CO_3⇄ CO_2 + H_2O
4.3.1.1	L-Aspartate ammonia-lyase	Aspartate ammonia-lyase	L-Aspartate$^+$⇄ Fumarate+NH_3
4.3.2.1	L-Argininosuccinate arginine-lyase	Argininosuccinate lyase	L-Argininosuccinate⇄ Fumarate+L-arginine
4.99.1.1	Protoheme ferro-lyase	Ferrochelatase	Protoporphyrin+Fe^{2+}⇄ Protoheme+$2H^+$
5. Isomerases: Intramolecular rearrangements			
5.1.1.4	Proline racemase	Proline racemase	L-Proline⇄ D-Proline
5.1.3.2	UDP-glucose 4-epimerase	UDP-glucose epimerase	UDP-glucose⇄ UDP-galactose
5.3.1.1	D-Glyceraldehyde-3-phosphate ketol-isomerase	Triosephosphate isomerase	D-Glyceraldehyde 3-phosphate⇄ Dihydroxyacetone phosphate
5.3.1.5	D-xylose ketol-isomerase	Xylose isomerase	D-Xylose⇄ D-Xylulose
5.4.99.2	Methylmalonyl-CoA CoA-carbonylmutase	Methylmalonyl-CoA mutase	Methylmalonyl-CoA⇄ Succinyl-CoA
6. Ligases (synthetases): Joining of molecules by covalent bond formation			
6.1.1.1	L-Tyrosine:tRNA ligase (AMP)	Tyrosyl-tRNA synthetase	ATP + L-Tyrosine+tRNA⇄ AMP + Pyrophosphate+L-Tyrosyl-tRNA
6.3.1.1	L-Aspartate:Ammonia ligase (AMP)	Asparagine synthetase	ATP + L-Aspartate+NH_3⇄ AMP + Pyrophosphate+L-Asparagine
6.3.1.2	L-Glutamate:Ammonia ligase (ADP)	Glutamine synthetase	ATP + L-Glutamate+NH_3⇄ ADP + Orthophosphate+L-Glutamine
6.3.4.5	L-Citrulline:L-Aspartate ligase (AMP)	Argininosuccinate synthetase	ATP + L-Citrulline+L-Aspartate⇄ AMP+ Pyrophosphate+L-Argininosuccinate
6.4.1.2	Acetyl-CoA: Carbon dioxide ligase (ADP)	Acetyl-CoA carboxylase	ATP + Pyruvate+CO_2 + H_2O⇄ ADP+ Orthophosphate+Malonyl-CoA

Enzyme examples selected in this table are those often referred to in this book

Table 3.2 Databases with enzyme EC numbers

Database	Website
BRENDA (The Comprehensive Enzyme Information System)	http://www.brenda-enzymes.info/
ExPASy (Enzyme nomenclature Database)	http://www.expasy.ch/enzyme/
KEGG (Kyoto Encyclopedia of Genes and Genomes)	http://www.genome.ad.jp/kegg/kegg2.html
MetaCyc (Pathway/Genome Databases)	http://metacyc.org/ and http://biocyc.org/
SYSTERS (Protein Family Database)	http://systers.molgen.mpg.de/
InterPro (database of protein families, domains, and functional sites)	http://www.ebi.ac.uk/interpro/
Protein Mutant Database	http://pmd.ddbj.nig.ac.jp/
BioCarta (Pathways of Life)	http://www.biocarta.com/
ExplorEnz (The Enzyme Database)	https://www.enzyme-database.org/
IUBMB (Enzyme Nomenclature)	http://www.chem.qmul.ac.uk/iubmb/enzyme/

3.4 SOME CONCERNS

The EC system of classification and nomenclature was arrived from a broad consensus with clear emphasis on the total reaction in question. The systematic names in a given class may be based on a written reaction, even if only the reverse reaction is experimentally demonstrated. This has created some situations that are less than perfect. While all enzyme-catalyzed reactions are reversible in principle (at least micro-reversibility at the active site!), the reaction-based classification for the forward direction would not be the same as that for the reverse direction. This was recognized very early by JBS Haldane, and according to him, calculation from thermodynamic data shows that catalase may act in the direction of H2O2 synthesis only under an O2 pressure of many billions of atmosphere! It would, therefore, be perverse, if logical, to describe it as water oxidase or for that matter the peroxidase as water dehydrogenase. To address such issues, the Commission has recommended that the more important direction of the overall reaction, from a biochemical view point, be used. It may be noted from Table 3.1 that reaction involving interconversion of NADH and NAD+ is all written in the direction where NAD+ is reduced by the other substrate. Also, when an overall reaction involves two types of reactions, then the second function is indicated in brackets. For instance, an oxidoreductase (decarboxylating) means the enzyme catalyzes an oxidation-reduction reaction in which one of the substrates is being decarboxylated.

Apart from issues related to the direction of overall reaction considered, a major concern is that the reaction mechanism is given less importance or completely ignored! Functionally distinct reactions are catalyzed by the following four enzymes:

Fumarate + H_2O ⇄ Malate	(Fumarate hydratase; EC 4.2.1.2)
Fumarate + NH_3 ⇄ Aspartate	(Aspartate ammonia-lyase; EC 4.3.1.1)
Fumarate + Arginine ⇄ Argininosuccinate	(Argininosuccinate lyase; EC 4.3.2.1)
Fumarate + AMP ⇄ Adenylosuccinate	(Adenylosuccinate lyase; EC 4.3.2.2)

All four enzymes are part of an evolutionary super-family of proteins; mechanistically they add elements of H-X to fumarate, using Michael reaction. This is a typical case where EC classification matches both in terms of overall reaction and the reaction mechanism. While UDP-glucose epimerase – based on the overall reaction – rightfully belongs to "isomerases" (Class 5), its intimate reaction mechanism tells a different story. The enzyme has tightly bound NAD+, and the reaction at C4 of the hexose involves redox chemistry. The reaction mechanism of glucosamine 6-phosphate isomerase is similar to that of ketose-aldose isomerase: D-Glucosamine 6-phosphate þ H2O⇄D-Fructose 6-phosphate þ NH3 It was therefore formerly grouped under Class 5 (with the code EC 5.3.1.10) as an isomerase. Subsequently this reaction was recognized as a C-N bond hydrolysis and reassigned as glucosamine 6-phosphate deaminase (EC 3.5.99.6).

Multiple Enzyme Forms and Isozymes At times we find that the same enzyme activity in an organism (or different organisms) is displayed by different protein forms. Enzymes that catalyze the same overall reaction but follow different mechanistic paths are not uncommon. Some examples of different protein forms are listed below (Table 3.3).

When these multiple molecular forms are coded by different but related genes (having different primary structure – amino acid sequence!), they are termed isozymes. Different isozyme forms of an enzyme are easily distinguished through their characteristic electrophoretic mobilities. The muscle and the heart forms of lactate dehydrogenase are the best studied examples of isozymes. All isozymes are examples of multiple forms of an enzyme, but all multiple forms need not be isozymes.

Table 3.3 Enzyme examples with different mechanistic forms

Enzyme	EC number	Mechanistic difference
Ribonucleotide reductase	EC 1.17.4.1	An iron protein
	EC 1.17.4.2	Requires a cobamide coenzyme
Methionine synthase	EC 2.1.1.13	Contains cobalamin
	EC 2.1.1.14	Does not contain cobalamin
Proteases	EC 3.4.4.x	Serine, cysteine, carboxylate, or metal ion at active site
Histidine decarboxylase	EC 4.1.1.22	Pyridoxal-phosphate (mammalian)
		Pyruvoyl prosthetic group (bacterial)
Fructose bisphosphate aldolase	EC 4.1.2.13	Class I – Forms a protonated imine
		Class II – Zinc polarized (microbial)
Dehydroquinase	EC 4.2.1.10	Type I – Forms a protonated imine
		Type II – Forms an enolate intermediate

Multiple molecular forms of the same enzyme may also arise due to other reasons. They may occur as (a) interconvertible forms through covalent modifications (e.g., phosphorylation of glycogen phosphorylase and adenylylation of E. coli glutamine synthetase), (b) proteolytic variants (chymotrypsinogen and chymotrypsin), (c) different oligomeric states of the same monomer (e.g., bovine liver glutamate dehydrogenase and avian liver acetyl CoA carboxylase), and (d) distinct conformational states of the same enzyme protein (e.g., R and T states of aspartate carbamoyltransferase). Occurrence of multiple enzyme forms is often associated with their role in metabolic regulation.

All multiple enzyme forms share the same EC number and a formal name. However these names need to be suitably prefixed or suffixed to indicate the modification, organ source, or organelle source. Much effort goes into characterizing any new enzyme. Therefore great care must be taken to define what it does and not repeat the name or indicate another reaction. In the final outcome, one needs not get bogged down by issues of semantics, for the excitement of enzymology beckons the uninitiated and the specialist alike.

Chapter 4

STRUCTURAL COMPONENTS OF ENZYMES

In this chapter we introduce the specific molecular components of enzymes that bring these forces to bear on the reactants and products of the catalyzed reaction. Like all proteins, enzymes are composed mainly of the 20 naturally occurring amino acids. We shall discuss how these amino acids link together to form the polypeptide backbone of proteins, and how these macromolecules fold to form the three-dimensional conformations of enzymes that facilitate catalysis. Individual amino acid side chains supply chemical reactivities of different types that are exploited by the enzyme in catalyzing specific chemical transformations. In addition to the amino acids, many enzymes utilize nonprotein *cofactors* to add additional chemical reactivities to their repertoire. We shall describe some of the more common cofactors found in enzymes, and discuss how they are utilized in catalysis.

4.1 THE AMINO ACIDS

An amino acid is any molecule that conforms, at neural pH, to the general formula:

$$H_3N^+\text{—}CH(R)\text{—}COO^-$$

The central carbon atom in this structure is referred to as the *alpha carbon* (Cα), and the substituent, R, is known as the *amino acid side chain*. Of all the possible chemical entities that could be classified as amino acids, nature has chosen to use 20 as the most common building blocks for constructing proteins and peptides. The structures of the side chains of the 20 naturally occurring amino acids are illustrated in Figure 4.1, and some of the physical properties of these molecules are summarized in Table 4.1. Since the alpha carbon is a chiral center, all the naturally occurring amino acids, except glycine, exist in two enantiomeric forms, ッ and セ. All naturally occurring proteins are composed exclusively from the ッ enantiomers of the amino acids.

As we shall see later in this chapter, most of the amino acids in a protein or peptide have their charged amino and carboxylate groups neutralized through peptide bond formation (in this situation the amino acid structure that remains is referred to as an amino acid *residue* of the protein or peptide). Hence, what chemically and physically distinguishes one amino acid from another in a protein is the identity of the side chain of the amino acid. As seen in Figure 4.1, these side chains vary in chemical structure from simple substituents, like a proton in the case of glycine, to complex bicyclic ring systems in the case of tryptophan. These different chemical structures of the side chains impart vastly different chemical reactivities to the amino acids of a protein. Let us review some of the chemical properties of the amino acid side chains that can participate in the interaction of proteins with other molecular and macromolecular species.

4.1.1 Properties of Amino Acid Side Chains

4.1.1.1 Hydrophobicity Scanning Figure 4.1, we note that several of the amino acids (valine, leucine, alanine, etc.) are composed entirely of hydrocarbons. One would expect that solvation of such nonpolar amino acids in a polar solvent like water would be thermodynamically costly. In general, when hydrophobic molecules are dissolved in a polar solvent, they tend to cluster together to minimize the amount of surface areas exposed to the solvent; this phenomenon is known as *hydrophobic attraction*. The repulsion from water of amino acids in a protein provides a strong driving force for proteins to fold into three-dimensional forms that sequester the nonpolar amino acids within the interior, or *hydrophobic core*, of the protein. Hydrophobic amino acids also help to stabilize the binding of nonpolar substrate molecules in the binding pockets of enzymes.

The hydrophobicity of the amino acids is measured by their tendency to partition into a polar solvent in mixed solvent systems. For example, a molecule can be dissolved in a 50 : 50 mixture of water and octanol. After mixing, the polar and nonpolar solvents are allowed to separate, and the concentration of the test molecule in each phase is measured. The equilibrium constant for transfer of the molecule from octanol to water is then given by:

$$K_{transfer} = \frac{[C_{H_2O}]}{[C_{oct}]} \quad (4.1)$$

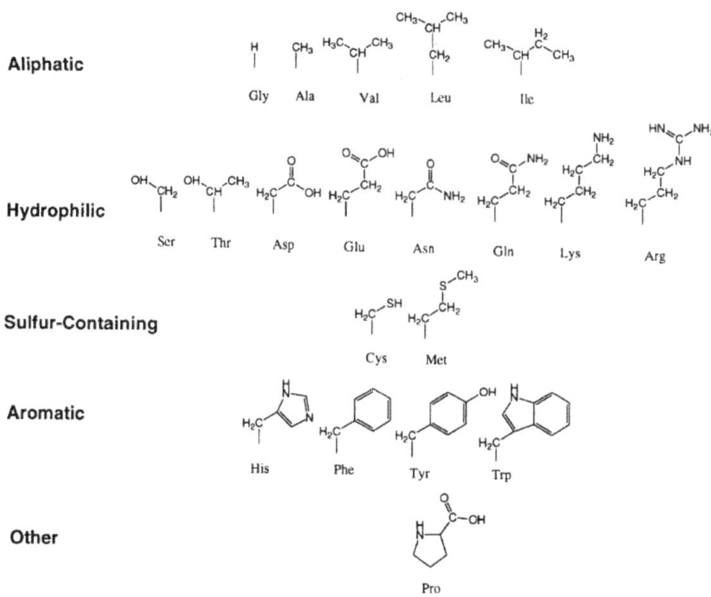

Figure 4.1 Side chain structures of the 20 natural amino acids. The entire proline molecule is shown.

where [CH2O] and [Coct] are the molar concentrations of the molecule in the aqueous and octanol phases, respectively. The free energy of transfer can then be calculated from the $K_{tranfer}$ value . Such thermodynamic studies have been performed for the transfer of the naturally occurring amino acids from a number of nonpolar solvent to water. To make these measurements more representative of the hydrophobicities of amino acids within proteins, workers use analogues of the amino acids in which the amino and carboxylate charges are neutralized (e.g., using *N*-acetyl ethyl esters of the amino acids). Combining this type of thermodynamic information for the different solvent systems, one can develop a rank order of hydrophobicities for the 20 amino acids. A popular rank order used in this regard is that developed by Kyte and Doolittle ; the Kyte and Doolittle hydrophobicity indices for the amino acid are listed in Table 3.1. In general, hydrophobic amino acids are found on the interior of folded proteins, where they are shielded from the repulsive forces of the polar solvent, and polar amino acids tend to be found on the solvent-exposed surfaces of folded proteins.

Table 4.1 Physicochemical properties of the natural amino acids

Amino Acid	Three-Letter Code	One-Letter Code	Mass of Residue in Proteins[a]	Accessible Surface Area ($Å^2$)[b]	Hydro-phobicity[c]	pK_a of Ionizable Side Chain	Occurrence in Proteins (%)[d]	Relative Mutability[e]	Van der Waals Volume ($Å^3$)
Alanine	Ala	A	71.08	115	+1.8		9.0	100	67
Arginine	Arg	R	156.20	225	−4.5	12.5	4.7	65	148
Asparagine	Asn	N	114.11	160	−3.5		4.4	134	96
Aspartate	Asp	D	115.09	150	−3.5	3.9	5.5	106	91
Cysteine	Cys	C	103.14	135	+2.5	8.4	2.8	20	86
Glutamate	Glu	E	128.14	180	−3.5	4.1	3.9	102	109
Glutamine	Gln	Q	129.12	190	−3.5		6.2	93	114
Glycine	Gly	G	57.06	75	−0.4		7.5	49	48
Histidine	His	H	137.15	195	−3.2	6.0	2.1	66	118
Isoleucine	Ile	I	113.17	175	+4.5		4.6	96	124
Leucine	Leu	L	113.17	170	+3.8		7.5	40	124
Lysine	Lys	K	128.18	200	−3.9	10.8	7.0	56	135
Methionine	Met	M	131.21	185	+1.9		1.7	94	124
Phenylalanine	Phe	F	147.18	210	+2.8		3.5	41	135
Proline	Pro	P	97.12	145	−1.6		4.6	56	90
Serine	Ser	S	87.08	115	−0.8		7.1	120	73
Threonine	Thr	T	101.11	140	−0.7		6.0	97	93
Tryptophan	Trp	W	186.21	255	−0.9		1.1	18	163
Tyrosine	Tyr	Y	163.18	230	−1.3	10.1	3.5	41	141
Valine	Val	V	99.14	155	+4.2		6.9	74	105

4.1.1.2 Hydrogen Bonding

Associated with the heteroatoms of the side chains of several amino acids are exchangeable protons that can serve as hydrogen donors for H-bonding. Other amino acids can participate as H-bond acceptors through the lone pair electrons on heteroatoms of their side chains. Hydrogen bonding of amino acid side chains and polypeptide backbone groups can greatly stabilize protein structures, as we shall see later in this chapter. Additionally, hydrogen bonds can be formed between amino acid side chains and ligand (substrates, products, inhibitors, etc.) atoms and can contribute to the overall binding energy for the interactions of enzymes with such molecules. Side chains that are capable of acting as H-bond donors include tyrosine (-O-H), serine (-O-H), threonine (-O-H), tryptophan (-N-H), histidine (-N-H), and cysteine (_S_H). At low pH the side chains of glutamic and aspartic acid can also act as H-bond donors (-COO-H). Heteroatoms on the side chains of the following amino acids can serve as H-bond acceptors: tyrosine (-:O:-H), glutamic and aspartic acid (-COO-), serine and threonine (-:O:-H), histidine (N:), cysteine (-:S:-H), and methionine (:S:). Several of the amino acids can serve as both donors and acceptors, of H bonds, as illustrated for tyrosine in Figure 4.2.

4.1.1.3 Salt Bridge Formation

Noncovalent electrostatic interactions can occur between electonegative and electropositive species within proteins. Figure 4.3 illustrates the formation of such an electrostatic interaction between the side chains of a lysine residue on one polypeptide chain and a glutamic acid residue on another polypeptide. Because these interactions resemble the ionic interactions associated with small molecule salt formation, they are often referred to as *salt bridges*. Salt bridges can occur intramolecularly, between a charged amino acid side chain and other groups within the protein, or intermolecularly, between the amino acid side chain and charged groups on a ligand or other macromolecule. For example, many proteins that bind to nucleic acids derive a significant portion of their binding energy by forming electrostatic interactions between positively charged amino acid residues on their surfaces (usually lysine and arginine residues) and the negatively charged phosphate groups of the nucleic acid backbone.

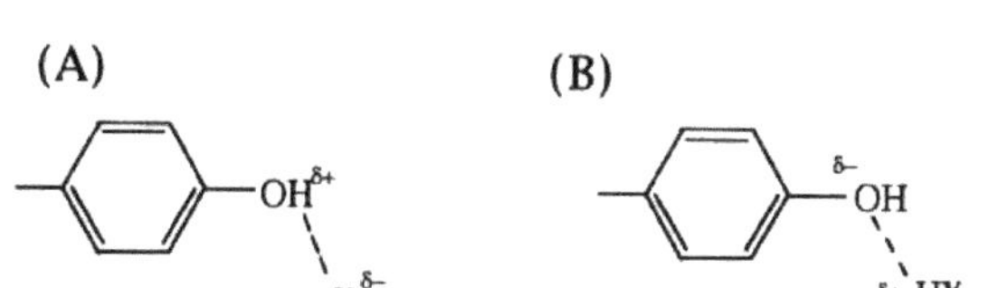

Figure 4.2 Tyrosine participation in hydrogen bonding as (A) a hydrogen donor and (B) a hydrogen acceptor.

Another example of the importance of these electrostatic interactions comes from the mitochondrial electron transfer cascade. Here electrons flow from the protein cytochrome *c* to the enzyme cytochrome oxidase, where they are used to reduce oxygen to water during cellular respiration. For the electron to jump from one protein to the other, the two must form a tight (dissociation constant≈10^{-9}M) complex. When the crystal structure of cytochrome *c* was solved, it became obvious that the surface of this molecule contained an area with an unusually high density of positively charged lysine residues. The putative binding site for cytochrome *c* on the cytochrome oxidase molecule has a corresponding high density of aspartic and glutamic acid residues. It is thus believed that the tight complex formed between these two proteins is facilitated by forming a large number of salt bridges at this interface. This suggestion is supported by the ability of the complex to be dissociated by adding high concentrations of salt to the solution. As the ionic strength increases, the salt ions compete for the counterions from, the amino acid residues that would otherwise participate in salt bridge formation.

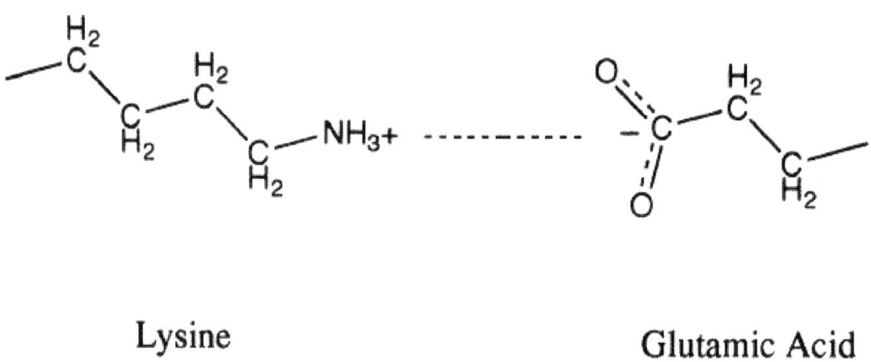

Figure 4.3 Salt bridge formation between a lysine and a glutamic acid residue at neutral pH.

4.1.2 Amino Acids as Acids and Bases

Surveying Table 4.1, we see that there are seven amino acid side chains, with titratable protons that can act as Brønsted—Lowry acids and conjugate bases. These are tyrosine, histidine, cysteine, lysine, and arginine, and aspartic and glutamic acids. The ability of these side chains to participate in acid—base chemistry provides enzymes with a mechanism for proton transfer to and from reactant and product molecules.

In addition to proton transfer, side chain Lewis acids and bases can participate in nucleophilic and electrophilic reactions with the reactant molecules, leading to bond cleavage and formation. The placement of acid and base groups from amino acid side chains, at critical positions within the active site, is a common mechanism exploited by enzymes to facilitate rapid chemical reactions with the molecules that bind in the active site. For example, hydrolysis of peptide and ester bonds can occur through nucleophilic attack of the peptide by water. This reaction goes through a transition state in which the carbonyl oxygen of the peptide has a partial negative charge, and the oxygen of water has a partial positive charge:

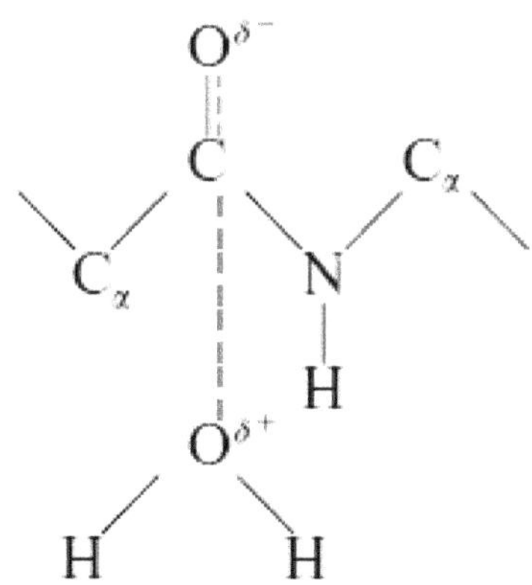

If one could place a basic group at a fixed position close to the water molecule, it would be possible to stabilize this transition state by partial transfer of one of the water protons to the base. This stabilization of the transition state would allow the reaction to proceed rapidly:

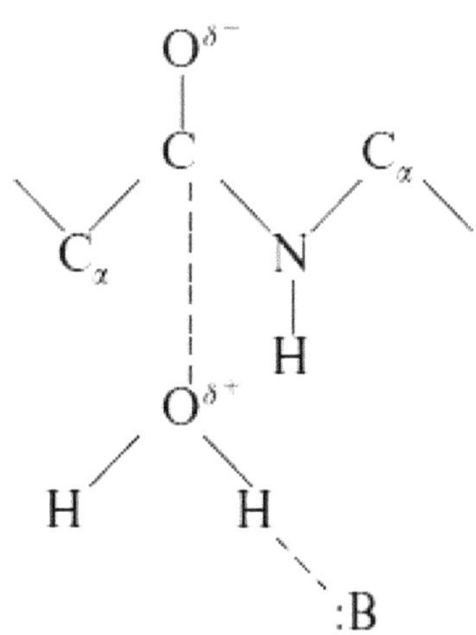

Alternatively, one could achieve the same stabilization by placing an acidic group at a fixed position in close proximity to the carbonyl oxygen, so that partial transfer of the proton from the acid to the carbonyl would stabilize the partial negative charge at this oxygen in the transition state. When one surveys the active sites of enzymes that catalyze peptide bond cleavage (a family of enzymes known as the proteases), one finds that there are usually acidic or basic amino acid side chains (or both) present at positions that are optimized for this type of transition state stabilization.

In discussing the acid and base character of amino acid side chains, it is important to recognize that the pK_a values listed in Table 4.1 are for the side chain groups in aqueous solution. In proteins, however, these pK_a values can be greatly affected by the local environment that is experienced by the amino acid residue. For example, the pK_a of glutamic acid in aqueous solution is 4.1, but the pK_a of particular glutamic acid residues in some proteins can be as high as 6.5. Thus, while the pK_a values listed in Table 4.1 can provide some insights into the probable roles of certain side chains in chemical reactions, caution must be exercised to avoid making oversimplifications.

4.1.3 Cation and Metal Binding

Many enzymes incorporate divalent cations (M^{2+}, Ca^{2+}, Zn^{2+}) and transition metal ions (Fe, Cu, Ni, Co, etc.) within their structures to stabilize the folded conformation of the protein or to make possible direct participation in the chemical reactions catalyzed by the enzyme. Metals can provide a template for protein folding, as in the zinc finger domain of nucleic acid binding proteins, the calcium ions of calmodulin, and the zinc structural center of insulin. Metal ions can also serve as redox centers for catalysis; examples include heme—iron centers, copper ions, and nonheme irons. Other metal ions can serve as electrophilic reactants in catalysis, as in the case of the active site zinc ions of the metalloproteases. Most commonly metals are bound to the protein portion of the enzyme by formation of coordinate bonds with certain amino acid side chains: histidine, tyrosine, cysteine, and methionine, and aspartic and glutamic acids. Examples of metal coordination by each of these side chains can be found in the protein literature.

The side chain imidazole ring of histidine is a particularly common metal coordinator. Histidine residues are almost always found in association with transition metal binding sites on proteins and are very often associated with divalent metal ion binding as well. Figure 4.4, for example, illustrates the coordination sphere of the active site zinc of the enzyme carbonic anhydrase. Zinc typically forms four coordinate bonds in a tetrahedral arrangement about the metal ion. In carbonic anhydrase, three of the four bonds are formed bycoordination to the side chains of histidine residues from the protein. The fourth coordination site is occupied by a water molecule that participates directly in catalysis. During the course of the enzyme-catalyzed reaction, the zinc—water bond is broken and replaced transiently by a bond between the metal and the carbon dioxide substrate of the reaction.

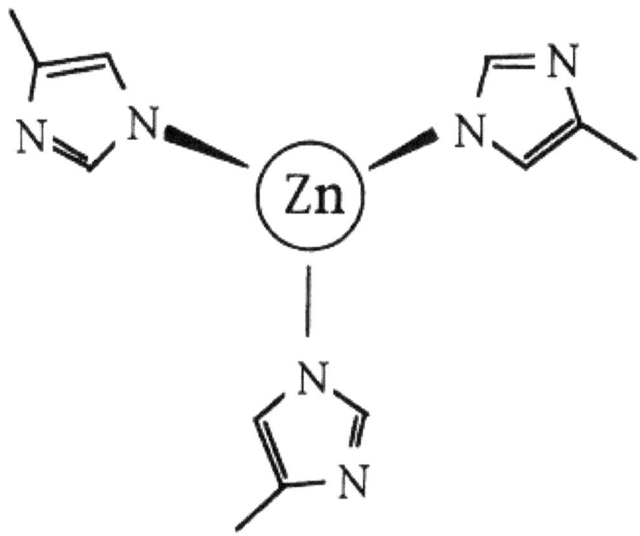

Figure 4.6 The coordination sphere of the active site zinc of carbonic anhydrase.

4.1.4 Anion and Polyanion Binding

The positively charged amino acids lysine and arginine can serve as counterions for anion and polyanion binding. Interactions of this type are important in binding of cofactors, reactants, and inhibitors to enzymes. Examples of anionic reactants and cofactors utilized by enzymes include phosphate groups, nucleotides and their analogues, nucleic acids, and heparin.

4.1.5 Covalent Bond Formation

We have pointed out that the chemical reactivities of amino acid residues within proteins are determined by the structures of their side chains. Several amino acids can undergo posttranslational modification (i.e., alterations that occur after the polypeptide chain has been synthesized at the ribosome) that alter their structure, hence reactivity, by means of covalent bond formation. In some cases, reversible modification of amino acid side chains is a critical step in the catalytic mechanism of the enzyme.

4.1.5.1 Disulfide Bonds

Two cysteine residues can cross-link, through an oxidative process, to form a sulfur—sulfur bond, referred to as a disulfide bond. These cross links can occur intramolecularly, between two cysteines within a single polypeptide, or intermolecularly, to join two polypeptides together. Such disulfide bond cross-linking can provide stabilizing energy to the folded conformation of the protein. Numerous examples exist of proteins that utilize both inter- and intramolecular disulfide bonds in their folded forms. Intermolecular disulfide bonds can also occur between a cysteine residue on a protein and a sulfhydryl group on a small molecule ligand or modifying reagent. For example, 4,4'--dithioldipyridine is a reagent used to quantify the number of free cysteines (those not involved in disulfide bonds) in proteins. The reagent reacts with the free sulfhydryls to form intermolecular disulfide bonds, with the liberation of a chromophoric by-product. The formation of the by-product is stoichiometric with reactive cysteines. Thus, one can quantify the number of cysteines that reacted from the absorbance of the by-product.

4.1.5.2 Phosphorylation

Certain amino acid side chains can be covalently modified by addition of a phosphate from inorganic phosphate (P). In nature, the phosphorylation of specific residues within proteins is facilitated by a class of enzymes known as the *kinases*. Another class of enzymes, the *phosphatases*, will selectively remove phosphate groups from these amino acids. This reversible phosphorylation/dephosphorylation can greatly affect the biological activity of enzymes, receptors, and proteins involved in protein—protein and protein—nucleic acid complex formation.

The most common sites for phosphorylation on proteins are the hydroxyl groups of threonine and serine residues; however, the side chains of tyrosine, histidine, and lysine can also be modified in this way (Figure 4.5). Tyrosine kinases, enzymes that specifically phosphorylate tyrosine residues within certain proteins, are of great current interest in biochemistry and cell biology. This is because it is recognized that tyrosine phosphorylation and dephosphorylation are critical in the transmission of chemical signals within cells (signal transduction).

Enzymes can also transiently form covalent bonds to phosphate groups during the course of catalytic turnover. In these cases, a *phosphoryl—enzyme intermediate* is formed by the transfer of an phosphate from substrate molecule or inorganic phosphate to specific amino acid side chains within the enzyme active site. Several examples of phosphoryl—enzyme intermediates are now known, which involve phosphoserine, phosphohistidine, and even phosphoaspartate formation. For example, the ATPases are enzymes that catalyze the hydrolysis of ATP (adenosine triphosphate) to form ADP (adenosine diphosphate) and P. In a subgroup of these enzymes, the Na^+, K^+, and the Ca^{2+}-ATPases, the γ-phosphate of ATP is transferred to the β-carboxylate of an aspartic acid residue of the enzyme during the reaction. Since the phosphoaspartate is thermodynamically unstable, it very quickly dissociates to liberate inorganic phosphate.

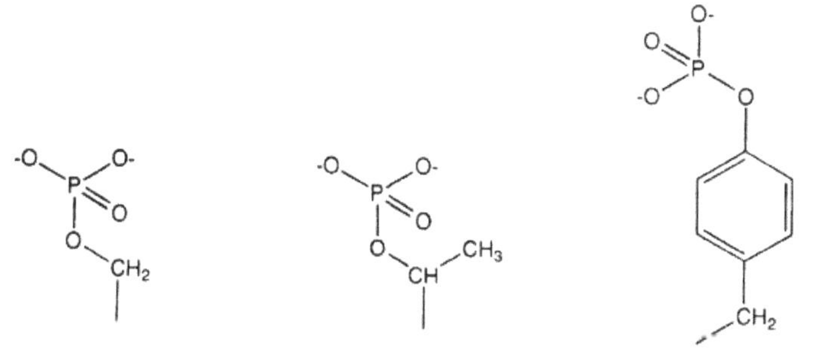

Phosphoserine Phosphothreonine Phosphotyrosine

Figure 4.5 The structures of the phosphorylated forms of serine, threonine, and tyrosine.

4.1.5.3 Glycosylation

In eukaryotic cells, sugars can attach to proteins by covalent bond formation at the hydroxyl groups of serine and threonine residues (O-linked glycosylation) or at the nitrogen of asparagine side chains (N-linked glycosylation). The resulting protein–sugar complex is referred to as a *glycoprotein*. The sugars used for this purpose are composed of monomeric units of galactose, glucose, manose, *N*-acetylglucosamine, *N*-acetylgalactosamine, sialic acid, fructose, and xylose. The presence of these sugar moieties can significantly affect the solubility, folding, and biological reactivity of proteins.

4.1.6 Steric Bulk

Aside from the chemical reactivities already discussed, the stereochemistry of the amino acid side chains plays an important role in protein folding and intermolecular interactions. The size and shape of the side chain determines the type of packing interactions that can occur with neighboring groups, according to their van der Waals radii. It is the packing of amino acid side chains within the active site of an enzyme molecule that gives overall size and shape to the binding cavity (pocket), which accommodates the substrate molecule; hence these packing interactions help determine the specificity for binding of substrate and inhibitor molecules at these sites. This is a critical aspect of enzyme catalysis; in Chapter 6 we shall discuss further the relationship between the size and shape of the enzyme binding pocket and the structure of ligands.

For the aliphatic amino acids, side chain surface area also influences the overall hydrophobicity of the residue. The hydrophobicity of aliphatic molecules, in general, has been correlated with their exposed surface area. Hansch and Coats have made the generalization that the $\Delta G_{transfer}$ from a nonpolar solvent, like *n*-octanol, to water increases by about 0.68 kcal/mol for every methylene group added to an aliphatic structure. While this is an oversimplification, it serves as a useful rule of thumb for predicting the relative hydrophobicities of structurally related molecules. This relationship between surface area and hydrophobicity holds not only for the amino acids that line the binding pocket of an enzyme, but also for the substrate and inhibitor molecules that might bind in that pocket.

4.2 THE PEPTIDE BOND

The peptide bond is the primary structural unit of the polypeptide chains of proteins. Peptide bonds result from the condensation of two amino acids, as follows:

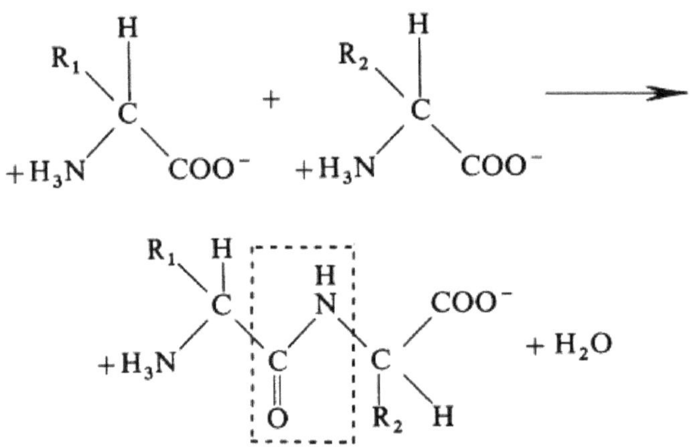

The product of such a condensation is referred to as a dipeptide, because it is composed of two amino acids. A third amino acid could condense with this dipeptide to form a tripeptide, a fourth to form a tetrapeptide, and so on. In this way chains of amino acids can be linked together to form *polypeptides* or *proteins.*

Until now we have drawn the peptide bond as an amide with double-bond character between the peptide carbon and the oxygen, and single-bond character between the peptide carbon and nitrogen atoms. Table 4.2 provides typical bond lengths for carbon—oxygen and carbon—nitrogen double and single bonds. Based on these data, one would expect the peptide carbon— oxygen bond length to be 1.22 Å and the carbon—nitogen bond length to be 1.45 Å. In fact, however, when x-ray crystallography was first applied to small peptides and other amide-containing molecules, it was found that the carbon— oxygen bond length was longer than expected, 1.24 Å, and the carbon— nitrogen bond length was shorter than expected, 1.32 Å (Figure 4.6). These values are intermediate between those expected for double and single bonds.

Table 4.2 Typical bond lengths for carbon--oxygen and carbon--nitrogen bonds

Bond Type	Bond Length (Å)
C—O	1.27
C=O	1.22
C—N	1.45
C=N	1.25

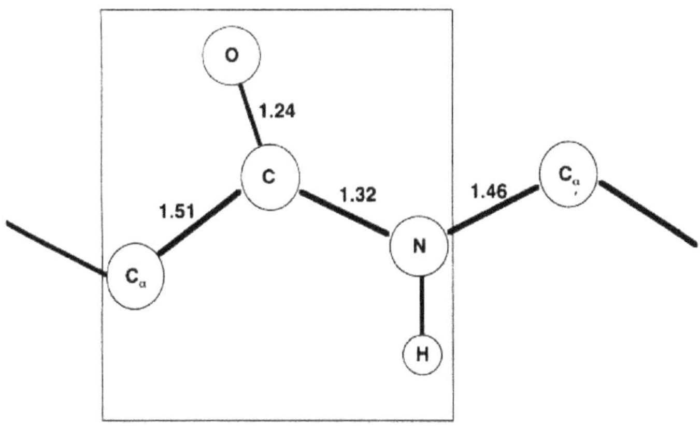

Figure 4.6 Schematic diagram of a typical peptide bond; numbers are typical bond lengths in,angstrom units.

These results were rationalized by the chemist Linus Pauling by invoking two resonance structures for the peptide bond; one as we have drawn it at the beginning of this section, with all the double-bond character on the carbon— oxygen bond, and another in which the double bond is between the carbon and the nitrogen, and the oxygen is negatively charged. Thus the - system is actually delocalized over all three atoms, O-C-N. Based on the observed bond lengths, it was concluded that a peptide bond has about 60% C=O character and about 40% C=N character.

The 40% double-bond character along the C—N axis results in about a 20 kcal/ mol resonance energy stabilization of the peptide. It also imposes a severe barrier to rotation about this axis. Hence, the six atoms associated with the peptide unit of polypeptides occur in a *planar* arrangement. The planarityof this peptide unit limits the configurations the polypeptide chain can adopt. It also allows for two stereoisomers of the peptide bond to occur: a *trans* configuration, in which the carbonyl oxygen and the nitrogenous proton are on opposite sides of the axis defined by the C-N bond, and a *cis* configuration, in which these groups are on the same side of this axis (Figure 4.7).

When proteins are produced on the ribosomes of cells, they are synthesized stereospecifically. They could be synthesized with either all-cis or all-trans peptide bonds. However, because of the steric bulk of the amino acid side chains, polypeptides composed of cis peptide bonds are greatly restricted in terms of the conformational space they can survey. Thus, there is a significant thermodynamic advantage to utilizing trans peptide bonds for proteins and unsurprisingly, almost all the peptide bonds in naturally occurring proteins are present in the trans configuration.

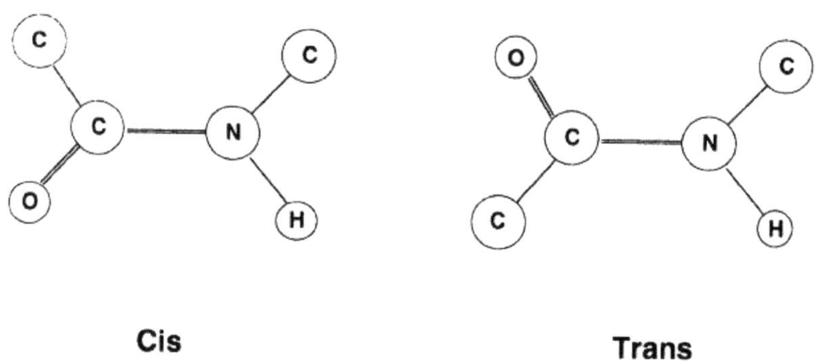

Figure 4.7 The cis and trans configurations of the peptide bond.

An exception to this general rule is found in prolyl—peptide bonds. Here the energy difference between the cis and trans isomers is much smaller (≈2 kcal/ mol), and so the cis isomer can occur without a significant disruption in stability of the protein. Nevertheless, only a very few examples of cis prolyl— peptide bonds have ever been observed in nature. Three examples are known of cis prolyl—peptide bonds within enzymes from x-ray crystallographic studies: in ribonuclease-S (before Pro 93 and before Pro 114), in a subtilisin (before Pro 168), and in staphylococcal nuclease (before Pro 116). Thus, while the cis prolyl—peptide bond is energetically feasible, it is extremely rare. For our purposes, then, we can assume that all the peptide bonds in the proteins we shall be discussing are present in the trans configuration.

4.3 AMINO ACID SEQUENCE OR PRIMARY STRUCTURE

The structure and reactivity of a protein are defined by the identity of the amino acids that make up its polypeptide chain, and by the order in which those amino acids occur in the chain. This information constitutes the *amino acid sequence* or *primary structure* of the protein. Recall that we can link amino acids together through condensation reactions to form polypeptide chains. We have seen that for most of the amino acids in such a chain, the condensation results in loss of the charged amino and carboxylate moieties. No matter how many times we perform this condensation reaction, however, the final polypeptide will always retain a charged amino group at one end of the chain and a charged carboxylate at the other end. The terminal amino acid that retains the positively charged amino group is referred to as the *N-terminus* or *amino terminus*. The other terminal amino acid, retaining the negatively charged carboxylate group, is referred to as the *C-terminus* or *carboxy terminus*.

The individual amino acids in a protein are identified numerically in sequential order, starting with the N-terminus. The N-terminal amino acid is labeled number 1, and the numbering continues in ascending numerical order, ending with the residue at the carboxy terminus. Thus, when we read in the literature about "active site residue Ser 530," it means that the 530th amino acid from the N-terminus of this enzyme is a serine, and it occurs within the active site of the folded protein.

With the advent of recombinant DNA technologies, it has become commonplace to substitute amino acid residues within proteins . One may read, for example, about a protein in which a His 323—Asn mutation was induced by means of site-directed mutagenesis. This means that in the natural, or *wild-type* protein, a histidine residue occupies position 323, but through mutagenesis this residue has been replaced by an asparagine to create a mutant (or altered) protein.

Studies in which mutant proteins have been purposely created through the methods of molecular biology are very common nowadays. It is important to remember, however, that mutations in protein sequences occur naturally as well. For the most part, point mutations of protein sequences occur with little effect on the biological activity of the protein, but in some cases the result is devastating. Consider, for example, the disease sickle cell anemia.

Patients with sickle cell anemia have a point mutation in their hemoglobin molecules. Hemoglobin is a tetrameric protein composed of four polypeptide chains: two identical chains, and two identical β chains. Together, the four polypeptides of a hemoglobin molecule contain about 600 amino acids. The β chains of normal hemoglobin have a glutamic acid residue at position 6. The crystal structure of hemoglobin reveals that residue 6 of the β chains is at the solvent-exposed surface of the protein molecule, and it is thus not surprising to find a highly polar side chain, like glutamic acid, at this position. In the hemoglobin from sickle cell anemia patients, this glutamic acid is replaced by a valine, a very nonpolar amino acid. When the hemoglobin molecule is devoid of bound oxygen (the deoxy form, which occurs after hemoglobin has released its oxygen supply to the muscles), these valine residues on different molecules of hemoglobin will come together to shield themselves from the polar solvent through protein aggregation. This aggregation leads to long fibers of hemoglobin in the red blood cells, causing the cells to adopt the narrow elongated "sickle" shape that is characteristic of this disease. Thus with only two amino acid changes out of 600 (one residue per β chain), the entire biological activity of the protein is severely altered. Sickle cell anemia was the first human disease that was shown to be caused by mutation of a specific protein (Pauling et al., 1949). We now know that there are many gene-based human diseases. Current efforts in "gene therapy" are aimed at correcting mutation-based diseases of these types.

4.4 SECONDARY STRUCTURE

We mentioned earlier that the delocalization of the peptide π system restricts rotation about the C-N bond axis. While this is true, it should be noted that free rotation is possible about the N-Cα and the Cα-C' bond axes (where C' represents the carbonyl carbon of the amino acid residue). The dihedral angles defined by these two rotations are represented by the symbols ϕ and ψ, respectively, and are illustrated for a dipeptide in Figure 4.8.

Upon surveying the observed values for ϕ and ψ for amino acid residues in the crystal structures of proteins, one finds that certain values occur with high frequency. For any amino acid except glycine, a plot of the observed ϕ and ψ angles looks like Figure 3.9. This type of graph is known as a Ramachandran plot, after the scientist who first measured these angles and constructed such plots. The most obvious feature of Figure 3.9 is that the values of ϕ and ψ tend to cluster around two pairs of angles. The two regions of high frequency correspond to the angles associated with two commonly found regular and repeating structural motifs within proteins, the *right-handed* α *helix* and the β-*pleated sheet*. Both these structural motifs are examples of protein *secondary structure*, an important aspect of the overall conformation of any protein.

4.4.1 The Right-Handed α Helix

The right-handed α helix is one of the most commonly found protein secondary structures (Figure 4.10). This structure was first predicted by Linus Pauling, on the basis of the stereochemical properties of polypeptides. The structure is stabilized by a network of hydrogen bonds between the carbonyl oxygen of one residue (i) and the nitrogenous proton of residue i+4. For most of the residues in the helix, there are thus two hydrogen bonds formed with neighboring peptide bonds, each contributing to the overall stability of the helix. As seen in Figure 4.10, this hydrogen-bonding network is possible because of the arrangement of C=O and N-H groups along the helical axis. The side chains of the amino acid residues all point away from the axis in this structure, thus minimizing steric crowding. The individual peptide bonds are aligned within the α-helical structure, producing, in addition, an overall dipole moment associated with the helix; this too adds some stabilization to the structure. The amino acid residues in an helix conform to a very precise stereochemical arrangement. Each turn of an helix requires 3.6 amino acid residues, with a translation along the helical axis of 1.5 Å per residue, or 5.4 Å per turn.

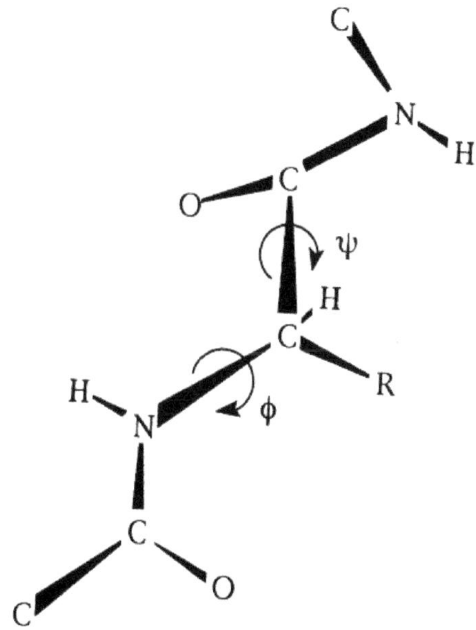

Figure 4.8 The dihedral angles of rotation for an amino acid in a peptide chain.

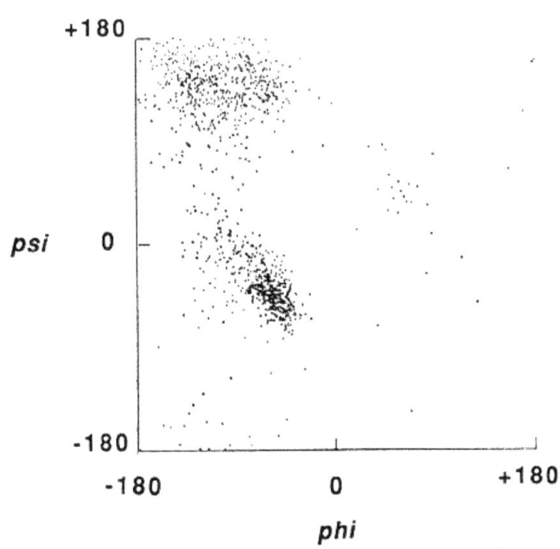

Figure 4.9 A Ramachandran plot for the amino acid alanine.

The network of hydrogen bonds formed in an α helix eliminates the possibility of those groups' participating in hydrogen bond formation with solvent—water molecules. For small peptides, the removal of these competing solvent—hydrogen bonds, by dissolving the peptide in aprotic solvents, tends to promote the formation of α helices. The same tendency is observed when regions of a polypeptide are embedded in the hydrophobic interior of a cell membrane. Within the membrane bilayer, where hydrogen bonding with solvent is not possible, peptides and polypeptides tend to form α-helical structures. The hydrocarbon core cross section of a typical biological membrane is about 30 Å. Since an helix translates 1.5 Å per residue, one can calculate that it takes, on average, about 20 amino acid residues, arranged in an α-helical structure, to traverse a membrane bilayer. For many proteins that are embedded in cell membranes (known as integral membrane proteins), one or more segments of 20 hydrophobic amino acids are threaded through the membrane bilayer as an α helix. These structures, often referred to as *transmembrane α helices*, are among the main mechanisms by which proteins associate with cell membranes.

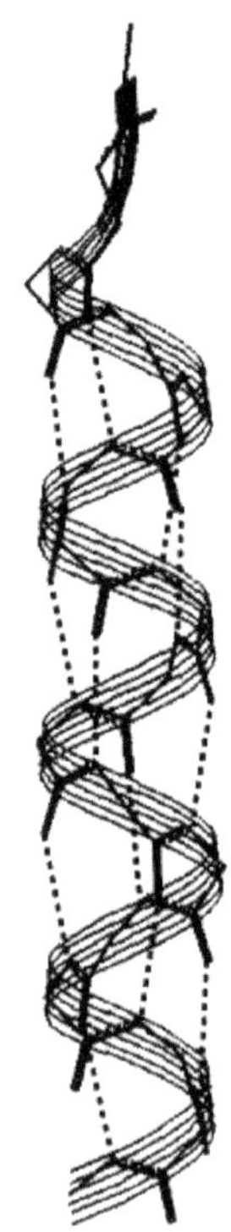

Figure 4.10 The right-handed α helix.

4.4.2 The β-Pleated Sheet

The β-pleated sheet, another very abundant secondary structure found in proteins, is composed of fully extended polypeptide chains linked together through interamide hydrogen bonding between adjacent strands of the sheet (Figure 4.11). Figure 4.11A illustrates a β-pleated sheet composed of two polypeptide chains. Note that the arrangement of peptide C=O and N-H groups permits this structure to be extended in either direction (i.e., to the left or right in the plane of the page as shown in Figure 4.11A) through the same type of interamide hydrogen bonding (Figure 4.11B).

For the moment, let us focus on a two-stranded βsheet, as in Figure 4.11A: the two component polypeptide chains could come from two distinct polypeptides (an *intermolecular* β *sheet*) or from two regions of the same contiguous polypeptide (an *intramolecular* β *sheet*); both types are found in natural proteins. If we imagine a β sheet within the plane of this page, we could have both chains running in the same direction, say from C-terminus at the top of the page to N-terminus at the botton. Alternatively, we could have the two chains running in opposite directions with respect to the placement of their Nand C-termini. These two situations describe structures referred to as *parallel* and *antiparallel* β-pleated sheets, respectively. Again, one finds both types in nature.

A

B

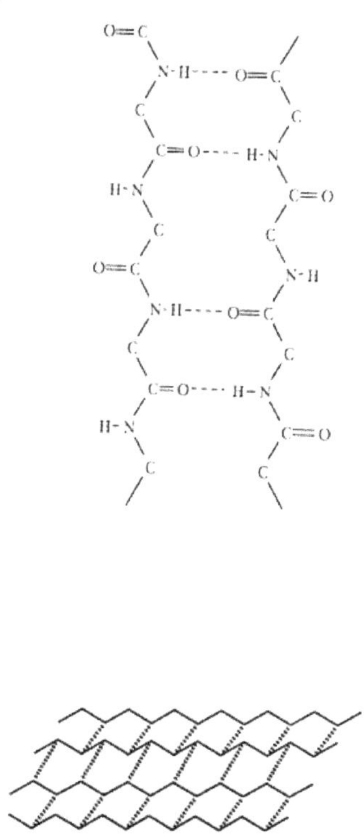

Figure 4.11 (A) A β-pleated sheet composed of two segments of polypeptide

held together by interchain hydrogen bonding. (B) An extended β-pleated sheet composed of four segments of polypeptide.

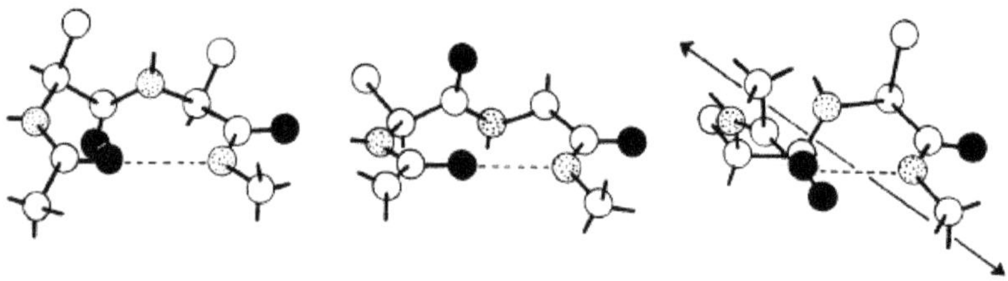

Figure 4.12 Three common forms of β turn.

4.4.3 β Turns

A third common secondary structure found in natural proteins is the β turn (also known as a reverse turn, hairpin turn, or β bend). The β turns are short segments of the polypeptide chain that allow it to change direction—that is, to *turn* upon itself. Turns are composed of four amino acid residues in a compact configuration in which an interamide hydrogen bond is formed between the first and fourth residue to stabilize the structure. Three types of β turn are commonly found in proteins: types I, II, and III (Figure 4.12).

Although turns represent small segments of the polypeptide chain, they occur often in a protein, allowing the molecule to adopt a compact three-dimensional structure. Consider, for example, an intramolecular antiparallel β sheet within a contiguous segment of a protein. To bring the two strands of the sheet into register for the correct hydrogen bonds to form, the polypeptide chain would have to change direction by 180°. This can be accomplished only by incorporating a type I or type II - turn into the polypeptide chain, between the two segments making up the - sheet. Thus - turns play a very important role in establishing the overall three-dimensional structure of a protein.

4.4.4 Other Secondary Structures

One can imagine other regular repeating structural motifs that are stereochemically possible for polypeptides. In a series of adjacent type III β turns, for example, the polypeptide chain would adopt a helical structure, different from the α helix, that is known as a 3_{10} helix. This structure is indeed found in proteins, but it is rare. Some proteins, composed of high percentages of a single amino acid type, can adopt specialized helical structures, such as the polyproline helices and polyglycine helices. Again, these are special cases, not commonly found in the vast majority of proteins.

Most proteins contain regions of well-defined secondary structures interspersed with segments of nonrepeating, unordered structure in a conformation commonly referred to as *random coil* structure. These regions provide dynamic flexibility to the protein, allowing it to change shape, or conformation. These structural fluctuations can play an important role in facilitating the biological activities of proteins in general. They have particular significance in the cycle of substrate binding, catalytic transformations, and product release that is required for enzymes to function.

4.5 TERTIARY STRUCTURE

The term "*tertiary structure*" refers to the arrangement of secondary structure elements and amino acid side chain interactions that define the three-dimensional structure of the folded protein. Imagine that a newly synthesized protein exists in nature as a fully extended polypeptide chain—it is said then to be *unfolded* (Figure 4.13A) [Actually there is debate over how fully extended the polypeptide chain really is in the unfolded state of a protein; some data suggest that even in the unfolded state, proteins retain a certain amount of structure.

Now suppose that this protein is placed under the set of conditions that will lead to the formation of elements of secondary structure at appropriate locations along the polypeptide chain (Figure 4.13B). Next, the individual elements of secondary structure arrange themselves in three-dimensional space, so that specific contacts are made between amino acid side chains and between backbone groups (Figure 4.13C). The resulting *folded* structure of the protein is referred to as its *tertiary structure*.

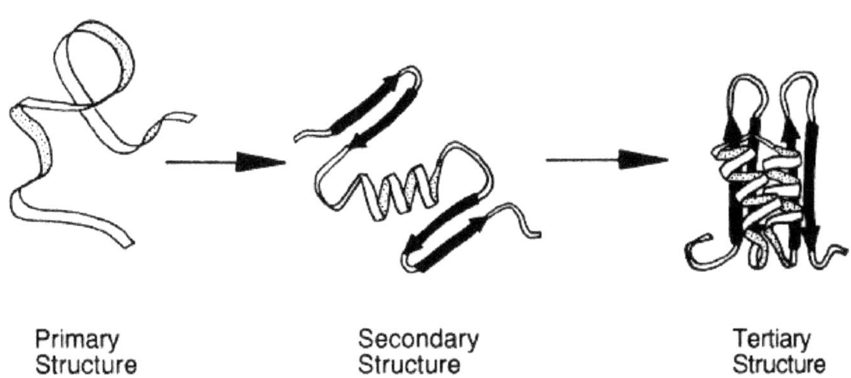

Figure 4.13 The folding of a polypeptide chain illustrating the hierarchy of protein structure from primary structure or amino acid sequence through secondary structure and tertiary structure.

What we have just described is the process of protein folding, which occurs naturally in cells as new proteins are synthesized at the ribosomes. The process is remarkable because under the right set of conditions it will also proceed spontaneously outside the cell in a test tube (in vitro). For example, at high concentrations chemicals like urea and guanidine hydrochloride will cause most proteins to adopt an unfolded conformation. In many cases, the subsequent removal of these chemicals (by dialysis, gel filtration chromatography, or dilution) will cause the protein to refold spontaneously into its correct native conformation (i.e., the folded state that occurs naturally and best facilitates the biological activity of the protein). The very ability to perform such experiments in the laboratory indicates that all the information required for the folding of a protein into its proper secondary and tertiary structures is encoded within the amino acid sequence of that protein.

Why is it that proteins fold into these tertiary structures? There are several important advantages to proper folding for a protein. First, folding provides a means of burying hydrophobic residues away from the polar solvent and exposing polar residues to solvent for favorable interactions. In fact, many scientists believe that the shielding of hydrophobic residues from the solvent is one of the strongest thermodynamic forces driving protein folding. Second, through folding the protein can bring together amino acid side chains that are distant from one another along the polypeptide chain. By bringing such groups into close proximity, the protein can form chemically reactive centers, such as the active sites of enzymes. An excellent example is provided by the serine protease chymotrypsin.

Serine proteases are a family of enzymes that cleave peptide bonds in proteins at specific amino acid residues . All these enzymes must have a serine residue within their active sites which functions as the primary nucleophile—that is, to attack the substrate peptide, thereby initiating catalysis. To enhance the nucleophilicity of this residue, the hydroxyl group of the serine side chain participates in hydrogen bonding with an active site histidine residue, which in turn may hydrogen-bond to an active site aspartate as shown in Figure 4.14. This "active site triad" of amino acids is a structural feature common to all serine proteases. In chymotrypsin this triad is composed of Asp 102, His 57, and Ser 195. As the numbering indicates, these three residues would be quite distant from one another along the fully extended polypeptide chain of chymotrypsin. However, the tertiary structure of chymotrypsin is such that when the protein is properly folded, these three residues come together to form the necessary interactions for effective catalysis. The tertiary structure of a protein will often provide folds or pockets within the protein structure that can accommodate small molecules. We have already used the term "active site" several times, referring, collectively, to the chemically reactive groups of the enzyme that facilitate catalysis. The active site of an enzyme is also defined by a cavity or pocket into which the substrate molecule binds to initiate the enzymatic reaction; the interior of this binding pocket is lined with the chemically reactive groups from the protein. As we shall see, there is a precise stereochemical relationship between the structure of the molecules that bind to the enzyme and that of the active site pocket. The same is generally true for the binding of agonists and antagonists to the binding pockets of protein receptors. In all these cases, the structure of the binding pocket is dictated by the tertiary structure of the protein.

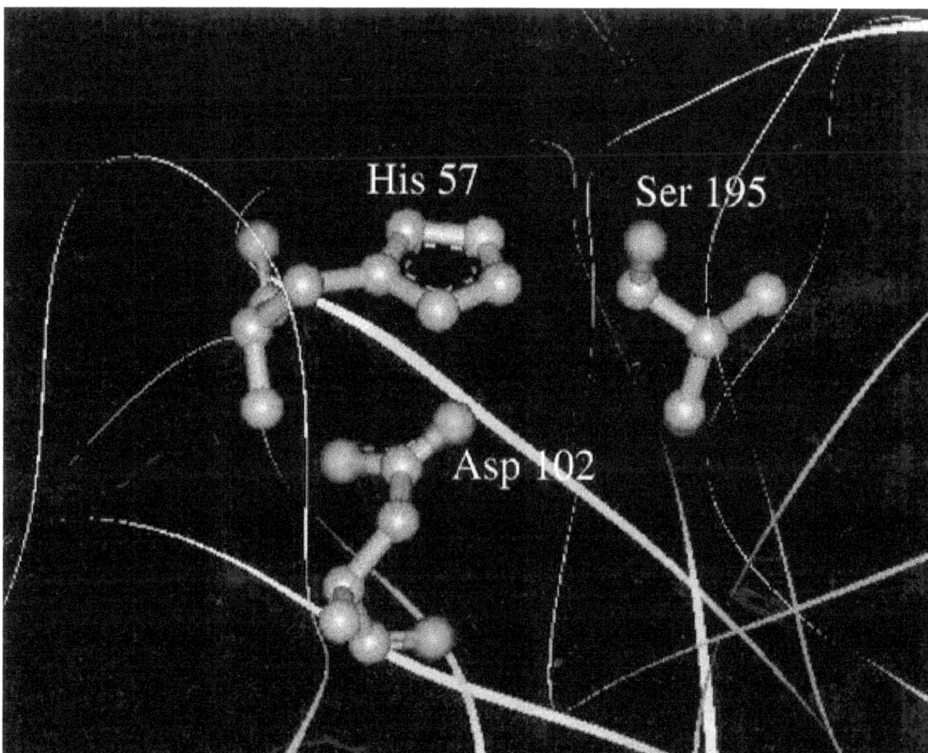

Figure 4.14 The active site triad of the serine protease α-chymotrypsin.

While no two proteins have completely identical three-dimensional structures, enzymes that carry out similar functions often adopt similar active site structures, and sometimes similar overall folding patterns. Some arrangements of secondary structure elements, which occur commonly in folded proteins, are referred to by some workers as *supersecondary structure*. Three examples of supersecondary structures are the helical bundle, the β barrel, and the β—α—β loop, illustrated in Figure 4.15.

In some proteins one finds discrete regions of compact tertiary structure that are separated by stretches of the polypeptide chain in a more flexible arrangement.

These discrete folded units are known as *domains*, and often they define functional units of the protein. For example, many cell membrane receptors play a role in signal transduction by binding extracellular ligands at the cell surface. In response to ligand binding, the receptor undergoes a structural change that results in macromolecular interactions between the receptor and other proteins within the cell cytosol. These interactions in turn set off a cascade of biochemical events that ultimately lead to some form of cellular response to ligand binding. To function in this capacity, such a receptor requires a minimum of three separate domains: an extracellular ligand binding domain, a transmembrane domain that anchors the protein within the cell membrane, and an intracellular domain that forms the locus for protein–protein interactions. These concepts are schematically illustrated in Figure 4.16.

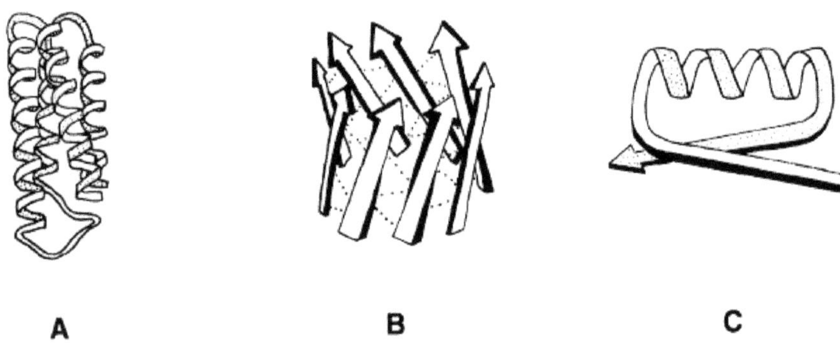

Figure 4.15 Examples of supersecondary structures: (A) a helical bundle, (B) a β barrel, and (C) a β–α–β loop.

Many enzymes are composed of discrete domains as well. For example, the crystal structure of the integral membrane enzyme prostaglandin synthase was recently solved by Garavito and his coworkers . The structure reveals three separate domains of the folded enzyme monomer: a β-sheet domain that functions as an interface for dimerization with another molecule of the enzyme, a membrane-incorporated α-helical domain that anchors the enzyme to the biological membrane, and a extramembrane globular (i.e., compact folded region) domain that contains the enzymatic active site and is thus the catalytic unit of the enzyme.

4.6 SUBUNITS AND QUATERNARY STRUCTURE

Not every protein functions as a single folded polypeptide chain. In many cases the biological activity of a protein requires two or more folded polypeptide chains to associate to form a functional molecule. In such cases the individual polypeptides of the active molecule are referred to as *subunits*. The subunits may be multiple copies of the same polypeptide chain (a homomultimer), or they may represent distinct polypeptides (a heteromultimer). In both cases the subunits fold as individual units, acquiring their own secondary and tertiary structures. The association between subunits may be stabilized through noncovalent forces, such as hydrogen bonding, salt bridge formation, and hydrophobic interactions, and may additionally include covalent disulfide bonding between cysteines on the different subunits.

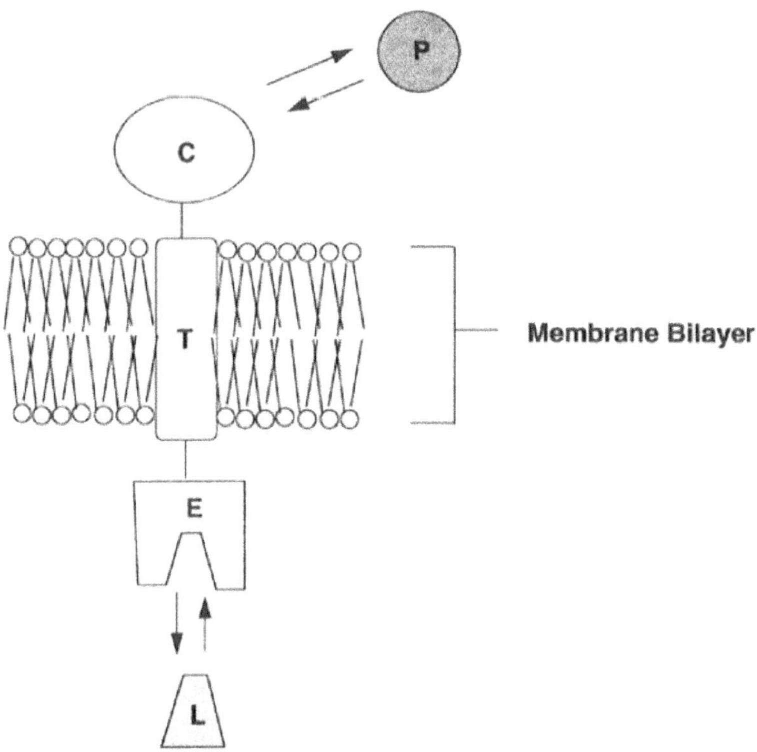

Figure 4.16 Cartoon illustration of the domains of a typical transmembrane receptor. The protein consists of three domains. The extracellular domain (E) forms the center for interaction with the receptor ligand (L). The transmembrane domain (T) anchors the receptor within the phospholipid bilayer of the cellular membrane. The cytosolic domain (C) extends into the intracellular space and forms a locus for interactions with other cytosolic proteins (P), which can then go on to transduce signals within the cell.

There are numerous examples of multisubunit enzymes in nature, and a few are listed in Table 4.3. In some cases, the subunits act as quasi-independent catalytic units. For example, the enzyme prostaglandin synthase exists as a homodimer, with each subunit containing an independent active site that processes substrate molecules to product. In other cases, the active site of the enzyme is contained within a single subunit, and the other subunits serve to stabilize the structure, or modify the reactivity of that active subunit. In the cytochrome oxidases, for example, all the active sites are contained in subunit I, and the other 3—12 subunits (depending of species) modify the stability and specific activity of subunit I. In still other cases the active site of the enzyme is formed by the coming together of the individual subunits. A good illustration of this comes from the aspartyl protease of the human immunodeficiency virus, HIV (the causal agent of AIDS). The active sites of all aspartyl proteases require a pair of aspartic acid residues for catalysis. The HIV protease is synthesized as a 99-residue polypeptide chain that dimerizes to form the active enzyme (a homodimer). Residue 25 of each HIV protease monomer is an aspartic acid residue. When the monomers combine to form the active homodimer, the two Asp 25 residues (designated Asp 25 and Asp 25' to denote their locations on separate polypeptide chains) come together to form the active site structure.

Table 4.3 Examples of multisubunit enzymes

Enzyme	Number of Subunits
HIV protease	2
Hexokinase	2
Bacterial cytochrome oxidase	3
Lactate dehydrogenase	4
Aspartate carbamoyl transferase	12
Human cytochrome oxidase	13

Without this subunit association, the enzyme could not perform its catalytic duties.

The arrangement of subunits of a protein relative to one another defines the *quaternary structure* of the protein. Consider a heterotrimeric protein composed of subunits A, B, and C. Each subunit folds into its own discrete tertiary structure. As suggested schematically in Figure 4.17, these three subunits could take up a number of different arrangements with respect to one another in three-dimensional space. This cartoon depicts two particular arrangements, or quaternary structures, that exist in equilibrium with each other. Changes in quaternary structure of this type can occur as part of the activity of many proteins, and these changes can have dramatic consequences.

An example of the importance of protein quaternary structure comes from examination of the biological activity of hemoglobin. Hemoglobin is the protein in blood that is responsible for transporting oxygen from the lungs to the muscles (as well as transporting carbon dioxide in the opposite direction). The active unit of hemoglobin is a heterotetramer, composed of two α subunits and two β subunits. Each of these four subunits contains a heme cofactor that is capable of binding a molecule of oxygen. The affinity of the heme for oxygen depends on the quaternary structure of the protein and on the state of oxygen binding of the heme groups in the other three subunits (a property known as cooperativity). Because of the cooperativity of oxygen binding to the hemes, hemoglobin molecules almost always have all four heme sites bound to oxygen (the oxy form) or all four heme sites free of oxygen (the deoxy form); intermediate forms with one, two, or three oxygen molecules bound are almost never observed.

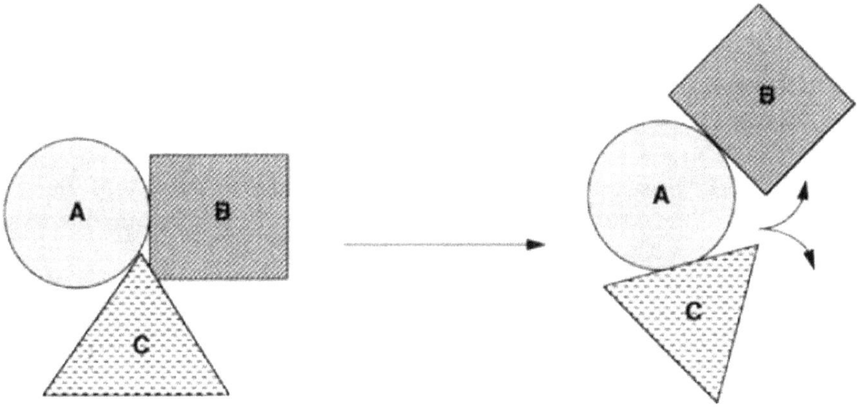

Figure 4.17 Cartoon illustrating the changes in subunit arrangements for a hypothetical heterotrimer that might result from a modification in quaternary structure.

When the crystal structures of oxy- and deoxyhemoglobin were solved, it was discovered that the two forms differed significantly in quaternary structure. If we label the four subunits of hemoglobin α_1, α_2, β_1, and β_2, we find that at the interface between the α_1 and β_2 subunits, oxygen binding causes changes in hydrogen bonding and salt bridges that lead to a compression of the overall size of the molecule, and a rotation of 15° for the $\alpha_1\beta_1$ pair of subunits relative to the $\alpha_2\beta_2$ pair (Figure 4.18). These changes in quaternary structure in part affect the relative affinity of the four heme groups for oxygen, providing a means of reversible oxygen binding by the protein. It is the reversibility of the oxygen binding of hemoglobin that allows it to function as a biological transporter of this important energy source; hemoglobin can bind oxygen tightly in the lungs and then release it in the muscles, thus facilitating cellular respiration in higher organisms.

4.7 COFACTORS IN ENZYMES

As we have seen, the structures of the 20 amino acid side chains can confer on enzymes a vast array of chemical reactivities. Often, however, the reactions catalyzed by enzymes require the incorporation of additional chemical groups to facilitate rapid reaction. Thus to fulfill reactivity needs that cannot be achieved with the amino acids alone, many enzymes incorporate nonprotein chemical groups into the structures of their active sites. These nonprotein chemical groups are collectively referred to as enzyme *cofactors* or *coenzymes*; Figure 4.19 presents the structures of some common enzyme cofactors.

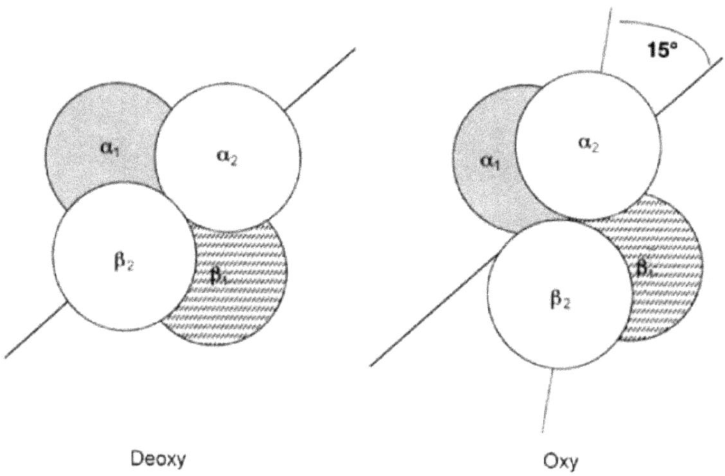

Figure 4.18 Cartoon illustration of the quaternary structure changes that accompany the binding of oxygen to hemoglobin.

In most cases, the cofactor and the enzyme associate through noncovalent interactions,(e.g., H-bonding, hydrophobic interactions). In some cases, however, the cofactors are covalently bonded to the polypeptide of the enzyme. For example, the heme group of the electron transfer protein cytochrome *c*, is bound to the protein through thioester bonds with two modified cysteine residues. Another example of covalent cofactor incorporation is the pyridoxal phosphate cofactor of the enzyme aspartate aminotransferase. Here the cofactor is covalently linked to the protein through formation of a Schiff base with a lysine residue in the active site.

In enzymes requiring a cofactor for activity, the protein portion of the active species is referred to as the *apoenzyme*, and the active complex between the protein and cofactor is called the *holoenzyme*. In some cases the cofactors can be removed to form the apoenzyme and be added back later to reconstitute the active holoenzyme. In some of these cases, chemically or isotopically modified versions of the cofactor can be incorporated into the apoenzyme to facilitate structural and mechanistic studies of the enzyme.

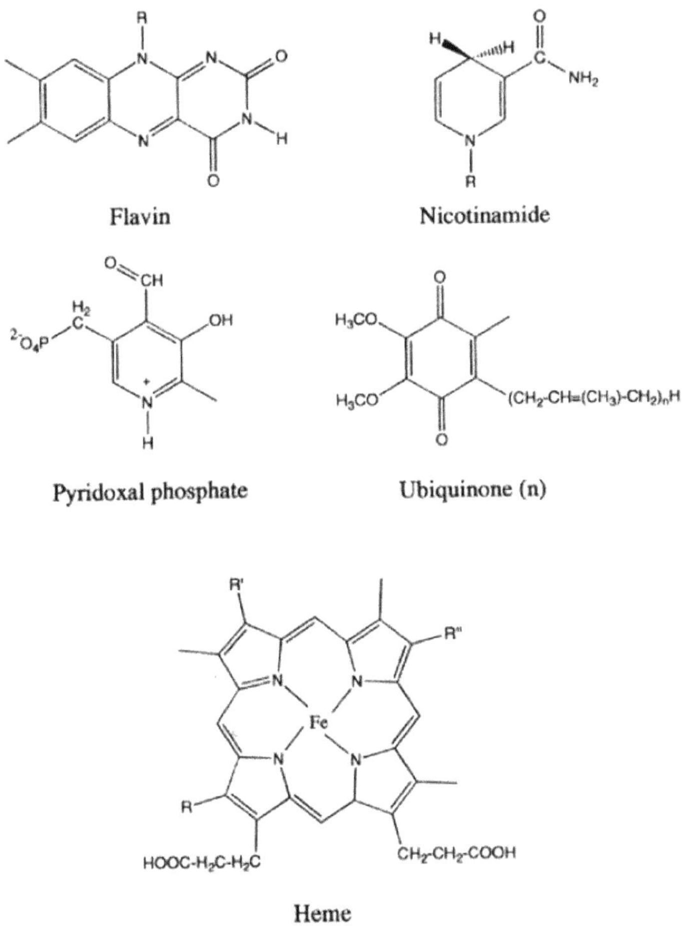

Figure 4.19 Examples of some common cofactors found in enzymes.

Cofactors fulfill a broad range of reactions in enzymes. One of the more common roles of enzyme cofactors is to provide a locus for oxidation/reduction (redox) chemistry at the active site. An illustrative example of this is the chemistry of flavin cofactors.

Flavins (from the Latin word *flavius*, meaning yellow) are bright yellow (λ_{max} =450nm) cofactors common to oxidoreductases, dehydrogenases, and electron transfer proteins. The main structural feature of the flavin cofactor is the highly conjugated isoalloxazine ring system (Figure 4.19). Oxidized flavins readily undergo reversible two-electron reduction to 1,5-dihydroflavin, and thus can act as electron sinks during redox reactions within the enzyme active site. For example, a number of dehydrogenases use flavin cofactors to accept two electrons during catalytic oxidation of NADH (another common enzyme cofactor; see Nicotinamide in Figure

4.19). Alternatively, flavins can undergo discrete one-electron reduction to form a semiquinone radical; this can be further reduced by a second one-electron reduction reaction to yield the fully reduced cofactor.

Through this chemistry, flavin cofactors can participate in one-electron oxidations, such as those carried out during respiratory electron transfer in mitochondria. There are actually two stable forms of the flavin semiquinone that interconvert, depending on pH (Figure 4.20) The blue neutral semiquinone occurs at neutral and acidic pH, while the red anionic semiquinone occurs above pH 8.4. Both forms can be stabilized and observed in certain enzymatic reactions.

Additional chemical versatility is demonstrated by flavin cofactors in their ability to form covalent adducts with substrate during redox reactions. The oxidation of dithiols to disulfides by the active site flavin of glutathione reductase is an example of this. Here the thiolate anion adds to the C4a carbon of the isoalloxazine ring system (Figure 4.21A). Likewise, in a number of flavoenzyme oxidases, catalytic reoxidation of reduced flavin by molecular oxygen proceeds with formation of a transient C4a peroxide intermediate (Figure 4.21B).

A variety of other cofactors participate in the catalytic chemistry of the enzyme active site. Some additional examples of these are listed in Table 4.4. This list is, however, far from comprehensive; rather it gives just a hint of the breadth of structures and reactivities provided to enzymes by various cofactors.

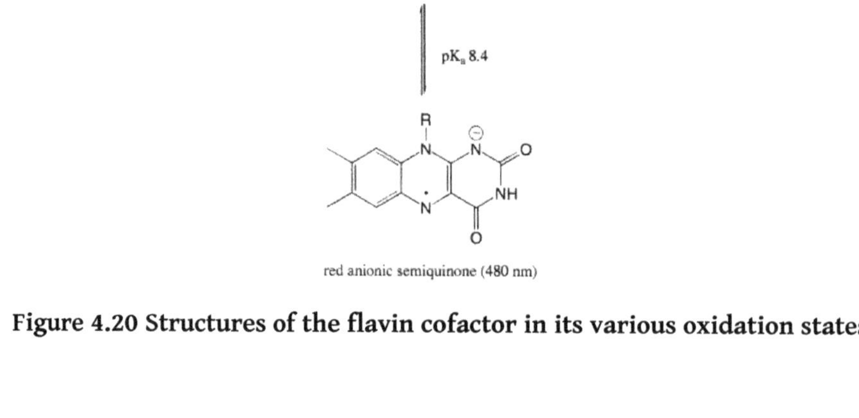

Figure 4.20 Structures of the flavin cofactor in its various oxidation states.

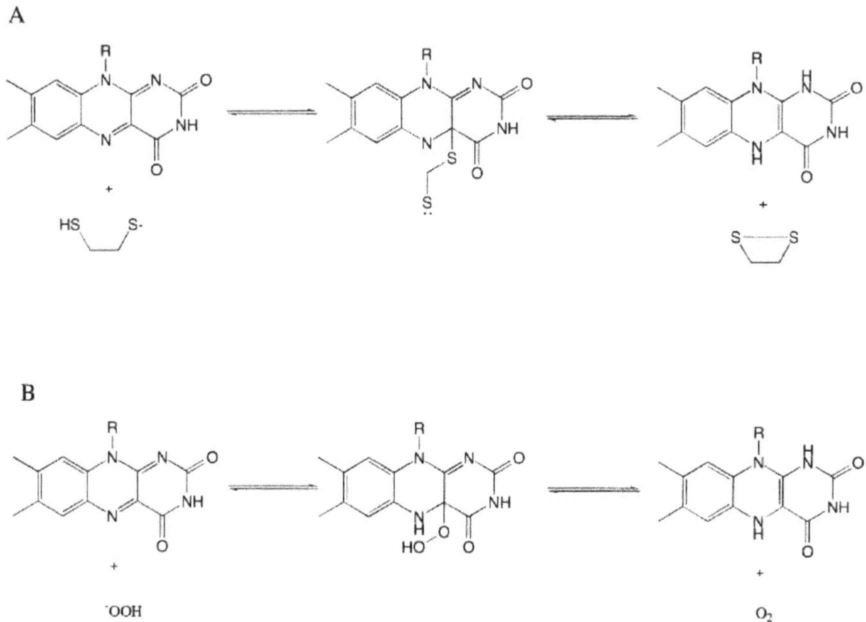

Figure 4.21 Covalent adduct formation by flavin cofactors in enzymes. (A) Oxidation of dithiols to disulfides through formation of a C4a–dithiol complex, as in the reaction of glutathione reductase with dithiols. (B) Reoxidation of reduced vin by molecular oxygen, with formation of a C4a–peroxide intermediate.

Table 4.4 Some examples of cofactors found in enzymes

Cofactor	Enzymatic Use	Examples of Enzyme
Copper ion	Redox center–ligand binding	Cytochrome oxidase, superoxide
Magnesium ion	Active site electrophile–phosphate binding	dismutase phosphodiesterases, ATP synthases
Zinc ion	Active site electrophile	Matrix metalloproteases, carboxypeptidase A
Flavins	Redox center–proton transfer	Glucose oxidase, succinate dehydrogenase
Hemes	Redox center–ligand binding	Cytochrome oxidase, cytochrome P450s
NAD and NADP	Redox center–proton transfer	Alcohol dehydrogenase, ornithine cyclase
Pyridoxal phosphate	Amino group transfer–stabilizer of intermediate carbanions	Aspartate transaminase, arginine racemase
Quinones	Redox center–hydrogen transfer	Cytochrome b_o, dihydroorotate dehydrogenase
Coenzyme A	Acyl group transfer	Pyruvate dehydrogenase,

Chapter 5

KINETICS OF SINGLE-SUBSTRATE ENZYME REACTIONS

Enzyme-catalyzed reactions can be studied in a variety of ways to explore different aspects of catalysis. Enzyme—substrate and enzyme—inhibitor complexes can be rapidly frozen and studied by spectroscopic means. Many enzymes have been crystallized and their structures determined by x-ray diffraction methods. More recently, enzyme structures have been determined by multidimensional NMR methods. Kinetic analysis of enzyme-catalyzed reactions, however, is the most commonly used means of elucidating enzyme mechanism and, especially when coupled with protein engineering, identifying catalytically relevant structural components. In this chapter we shall explore the use of steady state and transient enzyme kinetics as a means of defining the catalytic efficiency and substrate affinity of simple enzymes. As we shall see, the term *steady state* refers to experimental conditions in which the enzyme— substrate complex can build up to an appreciable "steady state" level. These conditions are easily obtained in the laboratory, and they allow for convenient interpretation of the time courses of enzyme reactions. All the data analysis described in this chapter rests on the ability of the scientist to conveniently measure the initial velocity of the enzyme-catalyzed reaction under a variety of conditions. For our discussion, we shall assume that some convenient method for determining the initial velocity of the reaction exists.

5.1 THE TIME COURSE OF ENZYMATIC REACTIONS

Upon mixing an enzyme with its substrate in solution and then (by some convenient means) measuring the amount of substrate remaining and/or the amount of product produced over time, one will observe *progress curves* similar to those shown in Figure 5.1. Note that the substrate depletion curve is the mirror image of the product appearance curve. At early times substrate loss and product appearance change rapidly with time but as time increases these rates diminish, reaching zero when all the substrate has been converted to product by the enzyme. Such time courses are well modeled by first-order kinetics,:

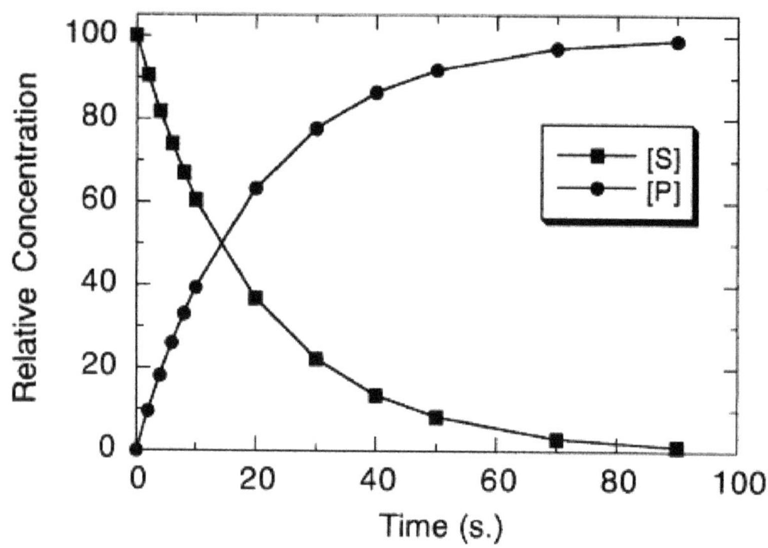

Figure 5.1 Reaction progress curves for the loss of substrate [S] and production of product [P] during an enzyme-catalyzed reaction.

$$[S] = [S_0]e^{-kt} \tag{5.1}$$

where [S] is the substrate concentration remaining at time t, $[S_0]$ is the starting substrate concentration, and k is the pseudo-first-order rate constant for the reaction. The velocity v of such a reaction is thus given by:

$$v = -\frac{d[S]}{dt} = \frac{d[P]}{dt} = k[S_0]e^{-kt} \tag{5.2}$$

Let us look more carefully at the product appearance profile for an enzymecatalyzed reaction (Figure 5.2). If we restrict our attention to the very early portion of this plot (shaded area), we see that the increase in product formation (and substrate depletion as well) tracks approximately linear with time. For this limited time period, the *initial velocity* v_0 can be approximated as the slope (change in *y* over change in *x*) of the linear plot of [S] or [P] as a function of time:

$$v_0 = -\frac{\Delta[S]}{\Delta t} = \frac{\Delta[P]}{\Delta t} \tag{5.3}$$

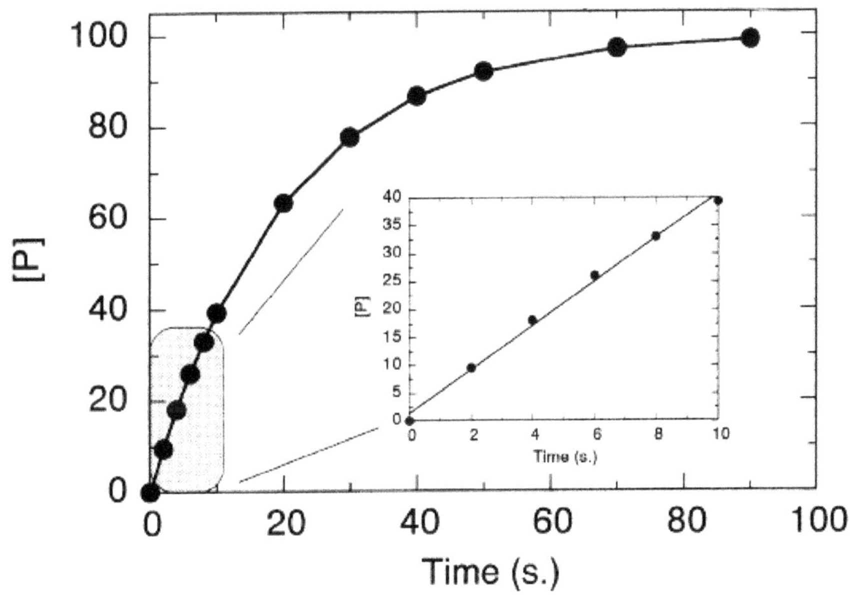

Figure 5.2 Reaction progress curve for the production of product during an enzyme-catalyzed reaction. Inset highlights the early time points at which the initial velocity can be determined from the slope of the linear plot of [P] versus time.

Experimentally one finds that the time course of product appearance and substrate depletion is well modeled by a linear function up to the time when about 10% of the initial substrate concentration has been converted to product . We shall see in next Chapter that by varying solution conditions, we can alter the length of time over which an enzyme-catalyzed reaction will display linear kinetics. For the rest of this chapter we shall assume that the reaction velocity is measured during this early phase of the reaction, which means that from here $v=v_0$, the initial velocity.

5.2 EFFECTS OF SUBSTRATE CONCENTRATION ON VELOCITY

From Equation 5.2, one would expect the velocity of a pseudo-first-order reaction to depend linearly on the initial substrate concentration. When early studies were performed on enzyme-catalyzed reactions, however, scientists found instead that the reactions followed the substrate dependence illustrated in Figure 5.3. Figure 5.3A illustrates the time course of the enzyme-catalyzed reaction observed at different starting concentrations of substrate; the velocities for each experiment are measured as the slopes of the plots of [P] versus time.

Figure 5.3B replots these data as the initial velocity v as a function of [S], the starting concentration of substrate. Rather than observing the linear relationship expected for first-order kinetics, we find the velocity apparently saturable at high substrate concentrations. This behavior puzzled early enzymologists.

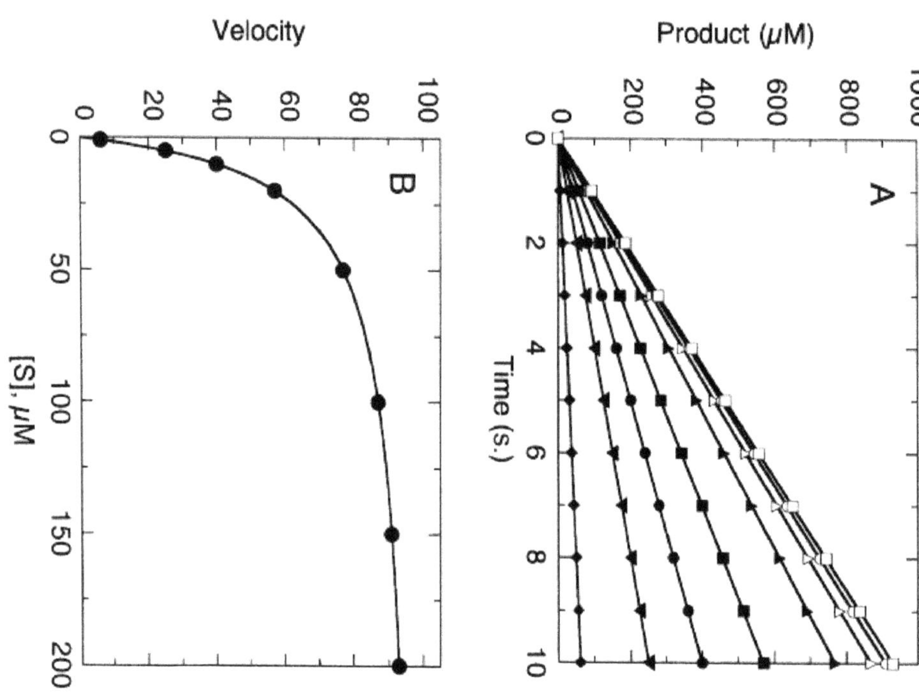

Figure 5.3 (A) Progress curves for a set of enzyme-catalyzed reactions with different starting concentrations of substrate [S]. (B) Plot of the reaction

velocities, measured as the slopes of the lines from (A), as a function of [S].

Three distinct regions of this curve can be identified: at low substrate concentrations the velocity appears to display first-order behavior, tracking linearly with substrate concentration; at very high concentrations of substrate, the velocity switches to zero-order behavior, displaying no dependence on substrate concentration; and in the intermediate region, the velocity displays a curvilinear dependence on substrate concentration. How can one rationalize these experimental observations?

A qualitative explanation for the substrate dependence of enzyme-catalyzed reaction velocities was provided by Brown (1902). At the same time that the kinetic characteristics of enzyme reactions were being explored, evidence for complex formation between enzymes and their substrates was also accumulating.

Brown thus argued that enzyme-catalyzed reactions could best be described by the following reaction scheme:

$$E + S \underset{k_{-1}}{\overset{k_1}{\rightleftharpoons}} ES \xrightarrow{k_2} E + P$$

This scheme predicts that the reaction velocity will be proportional to the concentration of the ES complex as: $v=k_2$ [ES]. Suppose that we held the total enzyme concentration constant at some low level and varied the concentration of S. At low concentrations of S the concentration of ES would be directly proportional to [S]; hence the velocity would depend on [S] in an apparent first-order fashion. At very high concentrations of S, however, practically all the enzyme would be present in the form of the ES complex. Under such conditions the velocity depends of the rate of the chemical transformations that convert ES to EP and the subsequent release of product to re-form free enzyme. Adding more substrate under these conditions would not effect a change in reaction velocity; hence the slope of the plot of velocity versus [S] would approach zero (as seen in Figure 5.3B). The complete [S] dependence of the reaction velocity (Figure 5.3B) predicted by the model of Brown resembles the results seen from the Langmuir isotherm Equation for equilibrium binding of ligands to receptors. This is not surprising, since in the model of Brown, catalysis is critically dependent on initial formation of a binary ES complex through equilibrium binding.

5.3 THE RAPID EQUILIBRIUM MODEL OF ENZYME KINETICS

Although the model of Brown provided a useful qualitative picture of enzyme reactions, to be fully utilized by experimental scientists, it needed to be put into a rigorous mathematical framework. This was accomplished first by Henri and subsequently by Michaelis and Menten . Ironically, Michaelis and Menten are more widely recognized for this contribution, although they themselves acknowledged the prior work of Henri. The basic rate equation derived in this section is commonly referred to as the Michaelis–Menten equation. Several writers have recently taken to referring to the equation as the Henri–Michaelis–Menten equation, in an attempt to correct this neglect of Henri's contributions. The reader should be aware, however, that the majority of the scientific literature continues to use the traditional terminology.

The Henri–Michaelis–Menten approach assumes that a rapid equilibrium is established between the reactants (E+S) and the ES complex, followed by slower conversion of the ES complex back to free enzyme and product(s); that is, this model assumes that $k_2 k_{-1}$ in the scheme presented in Section 5.2. In this model, the free enzyme E_f first combines with the substrate S to form the binary ES complex. Since substrate is present in large excess over enzyme, we can use the assumption that the free substrate concentration $[S]_f$ is well approximated by the total substrate concentration added to the reaction [S]. Hence, the equilibrium dissociation constant for this complex is given by:

$$K_S = \frac{[E]_f[S]}{[ES]} \tag{5.4}$$

Similar to the treatment of receptor–ligand binding , here the free enzyme concentration is given by the difference between the total enzyme concentration [E] and the concentration of the binary complex [ES]:

$$[E]_f = [E] - [ES] \tag{5.5}$$

and therefore,

$$K_S = \frac{([E] - [ES])[S]}{[ES]} \tag{5.6}$$

This can be rearranged to give an expression for [ES]:

$$[ES] = \frac{[E][S]}{K_S + [S]} \tag{5.7}$$

Next, the ES complex is transformed by various chemical steps to yield the product of the reaction and to recover the free enzyme. In the simplest case, a single chemical step, defined by the first-order rate constant k_2 , results in product formation. More likely, however, there will be a series of rapid chemical events following ES complex formation. For simplicity, the overall rate for these collective chemical steps can be described by a single first-order rate constant k_{cat}. Hence:

$$E + S \underset{K_S}{\rightleftharpoons} ES \xrightarrow{k_{cat}} E + P$$

and the rate of product formation is thus given by the first-order equation:

$$v = k_{cat}[ES] \tag{5.8}$$

Combining Equations 5.7 and 5.8, we obtain:

$$v = \frac{k_{cat}[E][S]}{K_S + [S]} \tag{5.9}$$

Equation 5.9 is similar to the equation for a Langmuir isotherm . This, then, describes the reaction velocity as a hyperbolic function of [S], with a maximum value of k_{cat}[E] at infinite [S]. We refer to this value as the maximum reaction velocity, or V_{max}.

$$V_{max} = k_{cat}[E] \tag{5.10}$$

Combining this definition with Equation 5.9, we obtain:

$$v = \frac{V_{max}[S]}{K_S + [S]} = \frac{V_{max}}{1 + \dfrac{K_S}{[S]}} \tag{5.11}$$

Equation 5.11 is the final equation derived independently by Henri and Michaelis and Menten to describe enzyme kinetic data. Note the striking similarity between this equation and the forms of the Langmuir isotherm equation presented in Chapter 3 (Equations 5.21 and 5.22). Thus, much of enzyme kinetics can be explained in terms of a simple equilibrium model involving rapid equilibrium between free enzyme and substrate to form the binary ES complex, followed by chemical transformation steps to produce and release product.

5.4 THE STEADY STATE MODEL OF ENZYME KINETICS

The original derivations by Henri and by Michaelis and Menten depended on a rapid equilibrium approach to enzyme reactions. This approach is quite useful in rapid kinetic measurements, such as single-turnover reactions, as described later in this chapter. The majority of experimental measurements of enzyme reactions, however, occur when the ES complex is present at a constant, steady state concentration (as defined below). Briggs and Haldane (1925) recognized that the equilibrium-binding approach of Henri and Michaelis and Menten could be described more generally by a steady state approach that did not require $k_2 < k_{-1}$

The following discussion is based on this description by Briggs and Haldane. As we shall see, the final equation that results from this treatment is very similar to Equation 5.11, and despite the differences between the rapid equilibrium and steady state approaches, the final steady state equation is commonly referred to as the Henri—Michaelis—Menten equation.

Steady state refers to a time period of the enzymatic reaction during which the rate of formation of the ES complex is exactly matched by its rate of decay to free enzyme and products. This kinetic phase can be attained when the concentration of substrate molecules is in great excess of the free enzyme concentration. To achieve a steady state, certain condition must be met, and these conditions allow us to make some reasonable assumption, which greatly simplify the mathematical treatment of the kinetics. These assumptions are as follows:

1. During the initial phase of the reaction progress curve (i.e., conditions under which we are measuring the linear initial velocity), there is no appreciable buildup of any intermediates other than the ES complex. Hence, all the enzyme molecules can be accounted for by either the free enzyme or by the enzyme—substrate complex. The total enzyme concentration [E] is therefore given by:

$$[E] = [E]_f + [ES] \tag{5.12}$$

2. As in the rapid equilibrium treatment, we assume that the enzyme is acting catalytically, so that it is present in very low concentration relative to substrate, that is, $[S]<[E]$. Hence, formation of the ES complex does not significantly diminish the concentration of free substrate. We can therefore make the approximation: $[S]_{f\sim}$ [S], where $[S]_f$ is the free substrate concentration and [S] is the total substrate concentration).
3. During the initial phase of the progress curve, very little product is formed relative to the total concentration of substrate. Hence, during this early phase [P]~0 and therefore depletion of [S] is minimal. At the initiation of the reaction there will be a rapid burst of formation of the ES complex followed by a kinetic phase in which the rate of formation of new ES complex is balanced by the rate of its decomposition back to free enzyme and product. In other words, during this phase the concentration of ES is constant. We refer to this kinetic phase as the *steady state*, which is defined by:

$$\frac{d[ES]}{dt} = 0 \tag{5.13}$$

Figure 5.4 illustrates the development and duration of the steady state for the enzyme cytochrome *c* oxidase interacting with its substrates cytochrome *c* and molecular oxygen. As soon as the substrates and enzyme are mixed, we see a rapid pre—steady state buildup of ES complex, followed by a long time window in which the concentration of ES does not change (the steady state phase), and finally a post—steady state phase characterized by significant depletion of the starting substrate concentration.

With these assumptions made, we can now work out an expression for the enzyme velocity under steady state conditions. As stated previously, for the simplest of reaction schemes, the pseudo-first-order progress curve for an enzymatic reaction can be described by:

$$v = k_2[\mathrm{ES}] \tag{5.14}$$

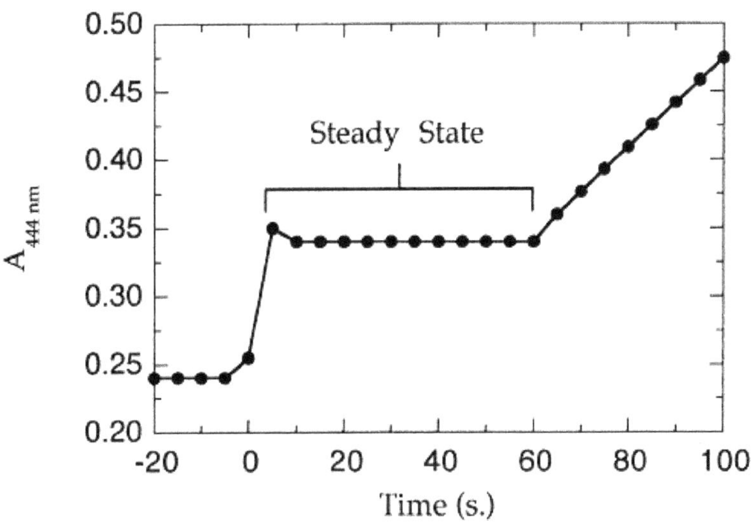

Figure 5.4 Development of the steady state for the reaction of cytochrome *c* oxidase with its substrates, cytochrome *c* and molecular oxygen. The absorbance at 444 nm reflects the ligation state of the active site heme cofactor of the enzyme. Prior to substrate addition (time-0) the heme group is in the Fe^{3+} oxidation state and is ligated by a histidine group from the enzyme. Upon substrate addition, the active site heme iron is reduced to the Fe^{2+} state and rapidly reaches a steady state phase of substrate utilization in which the iron is ligated by some oxygen species. The steady state phase ends when a significant portion of the molecular oxygen in solution has been used up. At this point the heme iron remains reduced (Fe^{2+}) but is no longer bound to a ligand at its sixth coordination site; this heme species has a much larger extinction coefficient at 444 nm; hence the rapid increase in absorbance at this wavelength following the steady state phase.

Now, [ES] is dependent on the rate of formation of the complex (governed by k_1) and the rate of loss of the complex (governed by $k_{-1}$and k_2). The rateequations for these two processes are thus given by:

$$\frac{d[\mathrm{ES}]}{dt} = k_1[\mathrm{E}]_f[\mathrm{S}]_f \qquad \text{and} \qquad \frac{-d[\mathrm{ES}]}{dt} = (k_{-1} + k_2)[\mathrm{ES}] \tag{5.15}$$

Under steady state conditions these two rates must be equal, hence:

$$k_1[\mathrm{E}]_f[\mathrm{S}]_f = (k_{-1} + k_2)[\mathrm{ES}] \tag{5.16}$$

This can be rearranged to obtain an expression for [ES]:

$$[\mathrm{ES}] = \frac{[\mathrm{E}]_f[\mathrm{S}]_f}{\dfrac{k_{-1} + k_2}{k_1}} \tag{5.17}$$

At this point let us define the term K_m as an abbreviation for the kinetic constants in the denominatior of the right-hand side of Equation 5.17:

$$K_m = \frac{k_{-1} + k_2}{k_1} \tag{5.18}$$

For now we will consider K_m to be merely an abbreviation to make our subsequent mathematical expressions less cumbersome. Later, however, we shall see that K_m has a more significant meaning. Substituting Equation 5.18 into Equation 5.17 we obtain:

$$[\mathrm{ES}] = \frac{[\mathrm{E}]_f[\mathrm{S}]_f}{K_m} \tag{5.19}$$

Now, since substrate depletion is insignificant during the steady state phase, we can replace the term $[S]_f$ by the total substrate concentration [S] (which is much more easily measured in real experimental situations). We can also use the equality of Equation 5.12 to replace $[E]_f$ by ([E]-[ES]). With these substitutions, Equation 5.19 can be recast as follows:

$$[\mathrm{ES}] = [\mathrm{E}]\frac{[\mathrm{S}]}{[\mathrm{S}] + K_m} \tag{5.20}$$

If we now combine this expression for [ES] with the velocity expression of Equation 5.14, we obtain:

$$v = k_2[\mathrm{E}]\frac{[\mathrm{S}]}{[\mathrm{S}] + K_m} \tag{5.21}$$

Or, we can generalize Equation 5.21 for more complex reaction schemes by substituting k_{cat} for k_2:

$$v = k_{cat}[E]\frac{[S]}{[S] + K_m} \quad (5.22)$$

As described earlier, as the concentration of substrate goes towards infinity, the velocity reaches a maximum value that we have defined as V_{max}. Under these conditions, the K_m term is a very small contribution to Equation 5.22. Therefore:

$$\lim_{[S]\to\infty} \frac{[S]}{[S] + K_m} \cong \frac{[S]}{[S]} = 1 \quad (5.23)$$

and thus we again arrive at Equation 5.10: $V_{max}=k_{cat}[E]$. Combining this with Equation 5.23 we finally arrive at an expression very similar to that first described by Henri and Michaelis and Menten (i.e., similar to Equation 5.11):

$$v = \frac{V_{max}[S]}{K_m + [S]} = \frac{V_{max}}{1 + \frac{K_m}{[S]}} \quad (5.24)$$

This is the central expression for steady state enzyme kinetics. While it differs from the equilibrium expression derived by Henri and by Michaelis and Menten, it is nevertheless universally referred to as the Michaelis—Menten or Henri—Michaelis—Menten equation.

In our definition of K_m (Equation 5.18), we combined first-order rate constants (k_{-1} and k_2, which have units of reciprocal time) with a second-order rate constant (k_1, which has units of reciprocal molarity, reciprocal time) in such a way that the resulting K_m has units of molarity, as does [S]. If we set up our experimental system so that the concentration of substrate exactly matches K_m, Equation 5.24 will reduce to:

$$v = \frac{V_{max}[S]}{[S] + [S]} = \frac{V_{max}}{2} \quad (5.25)$$

This provides us with a working definition of K_m: *The K_m is the substrate concentration that provides a reaction velocity that is half of the maximal velocity obtained under saturating substrate conditions.* The K_m value is often referred to in the literature as the *Michaelis constant.* In comparing Equation 5.24 for steady state kinetics with Equation 5.11 for the rapid equilibrium treatment, we see that the equations are identical except for the substitution of K_m for K_s in the steady state treatment. It is therefore easy to confuse these terms and to treat K_m as if it were the thermodynamic dissociation constant for the ES complex. However, the two constants are not always equal, even in considerations of the simplest of reactions schemes, as here. Recall that K_s can be defined by the rato of the reverse and forward reaction rate constants:

$$K_S = \frac{k_{-1}}{k_1} \tag{5.26}$$

This value is not identical to the expression for K_m given in Equation 5.18. Only under the specific conditions that $k_2 < k_{-1}$ are K_m and K_s equivalent. For more complex reaction schemes one would replace the k_2 term in Equation 5.18 by k_{cat}. Recall that k_{cat} reflects a summation of multiple chemical steps in catalysis. Hence, depending on the details of the reaction mechanism, and the values of the individual rate constants, situations can arise in which the value of K_m is less than, greater than, or equal to K_s. Therefore, K_m should generally be considered as a kinetic, not thermodynamic, constant.

5.5 THE SIGNIFICANCE OF KCAT AND KM

We have gone to great lengths in this chapter to define and derive expressions for the kinetic constants k_{cat} and K_m. What value do these constants add to our understanding of the enzyme under study?

5.5.1 Km

The value of K_m varies considerably from one enzyme to another, and for a particular enzyme with different substrates. We have already defined K_m as the substrate concentration that results in half-maximal velocity for the enzymatic reaction. An equivalent way of stating this is that the K_m represents the substrate concentration at which half of the enzyme active sites in the sample are filled (i.e., saturated) by substrate molecules in the steady state. Hence, while K_m is not equivalent to K_m under most conditions, it can nevertheless be used as a relative measure of substrate binding affinity. In some instances, changes in solution conditions (pH, temperature, etc.) can have selective effects on the value of K_m. Also, one sometimes observes effects on the value of K_m in the course of comparing different mutants or isoforms of an enzyme, or different substrates with a common enzyme. In these cases one can reasonably relate the changes to effects on the stability (i.e., affinity) of the ES complex. As we shall see below, however, the ratio k_{cat}/K_m is generally a better measure of effects on substrate binding.

5.5.2 kcat

Considering Equations 5.22—5.24, we see that if one knows the concentration of enzyme used experimentally, the value of k_{cat} can be directly calculated by dividing the experimentally determined value of V_{max} by [E]. The value of k_{cat} is sometimes

Or, we can generalize Equation 5.21 for more complex reaction schemes by substituting k_{cat} for k_2:

$$v = k_{\text{cat}}[\text{E}]\frac{[\text{S}]}{[\text{S}] + K_{\text{m}}} \tag{5.22}$$

As described earlier, as the concentration of substrate goes towards infinity, the velocity reaches a maximum value that we have defined as V_{max}. Under these conditions, the K_m term is a very small contribution to Equation 5.22. Therefore:

$$\lim_{[\text{S}]\to\infty} \frac{[\text{S}]}{[\text{S}] + K_{\text{m}}} \cong \frac{[\text{S}]}{[\text{S}]} = 1 \tag{5.23}$$

and thus we again arrive at Equation 5.10: $V_{max}=k_{cat}[E]$. Combining this with Equation 5.23 we finally arrive at an expression very similar to that first described by Henri and Michaelis and Menten (i.e., similar to Equation 5.11):

$$v = \frac{V_{\text{max}}[\text{S}]}{K_{\text{m}} + [\text{S}]} = \frac{V_{\text{max}}}{1 + \frac{K_{\text{m}}}{[\text{S}]}} \tag{5.24}$$

This is the central expression for steady state enzyme kinetics. While it differs from the equilibrium expression derived by Henri and by Michaelis and Menten, it is nevertheless universally referred to as the Michaelis—Menten or Henri—Michaelis—Menten equation.

In our definition of K_m (Equation 5.18), we combined first-order rate constants (k_{-1} and k_2, which have units of reciprocal time) with a second-order rate constant (k_1, which has units of reciprocal molarity, reciprocal time) in such a way that the resulting K_m has units of molarity, as does [S]. If we set up our experimental system so that the concentration of substrate exactly matches K_m, Equation 5.24 will reduce to:

$$v = \frac{V_{\text{max}}[\text{S}]}{[\text{S}] + [\text{S}]} = \frac{V_{\text{max}}}{2} \tag{5.25}$$

This provides us with a working definition of K_m: *The K_m is the substrate concentration that provides a reaction velocity that is half of the maximal velocity obtained under saturating substrate conditions.* The K_m value is often referred to in the literature as the *Michaelis constant.* In comparing Equation 5.24 for steady state kinetics with Equation 5.11 for the rapid equilibrium treatment, we see that the equations are identical except for the substitution of K_m for K_s in the steady state treatment. It is therefore easy to confuse these terms and to treat K_m as if it were the thermodynamic dissociation constant for the ES complex. However, the two constants are not always equal, even in considerations of the simplest of reactions schemes, as here. Recall that K_s can be defined by the rato of the reverse and forward reaction rate constants:

$$K_S = \frac{k_{-1}}{k_1} \tag{5.26}$$

This value is not identical to the expression for K_m given in Equation 5.18. Only under the specific conditions that $k_2 < k_{-1}$ are K_m and K_s equivalent. For more complex reaction schemes one would replace the k_2 term in Equation 5.18 by k_{cat}. Recall that k_{cat} reflects a summation of multiple chemical steps in catalysis. Hence, depending on the details of the reaction mechanism, and the values of the individual rate constants, situations can arise in which the value of K_m is less than, greater than, or equal to K_s. Therefore, K_m should generally be considered as a kinetic, not thermodynamic, constant.

5.5 THE SIGNIFICANCE OF KCAT AND KM

We have gone to great lengths in this chapter to define and derive expressions for the kinetic constants k_{cat} and K_m. What value do these constants add to our understanding of the enzyme under study?

5.5.1 Km

The value of K_m varies considerably from one enzyme to another, and for a particular enzyme with different substrates. We have already defined K_m as the substrate concentration that results in half-maximal velocity for the enzymatic reaction. An equivalent way of stating this is that the K_m represents the substrate concentration at which half of the enzyme active sites in the sample are filled (i.e., saturated) by substrate molecules in the steady state. Hence, while K_m is not equivalent to K_m under most conditions, it can nevertheless be used as a relative measure of substrate binding affinity. In some instances, changes in solution conditions (pH, temperature, etc.) can have selective effects on the value of K_m. Also, one sometimes observes effects on the value of K_m in the course of comparing different mutants or isoforms of an enzyme, or different substrates with a common enzyme. In these cases one can reasonably relate the changes to effects on the stability (i.e., affinity) of the ES complex. As we shall see below, however, the ratio k_{cat}/K_m is generally a better measure of effects on substrate binding.

5.5.2 kcat

Considering Equations 5.22—5.24, we see that if one knows the concentration of enzyme used experimentally, the value of k_{cat} can be directly calculated by dividing the experimentally determined value of V_{max} by [E]. The value of k_{cat} is sometimes referred to as the *turnover number* for the enzyme, since it defines the number of catalytic turnover events that occur per unit time. The units of k_{cat} are reciprocal time (e.g., min^{-1}, s^{-1}). Turnover numbers, however, are typically reported in units of molecules of product produced per unit time per molecules of enzyme present. As long as the same units are used to express the amount of product produced and the amount of enzyme present, these units will cancel and, as expected, the final units will be reciprocal time. It is important, however, that the units of product and enzyme concentration be expressed in molar or molarity units. In crude enzyme samples, such as cell lysates and other nonhomogeneous protein samples, it is often impossible to know the concentration of enzyme in anything other than units of total protein mass. The literature is thus filled with enzyme activity values expressed as number of micrograms of product produced per minute per microgram of protein in the enzyme sample. While such units may be useful in comparing one batch of crude enzyme to another , it is difficult to relate these values to kinetic constants, such as k_{cat}.

In the laboratory we can easily determine the turnover number as k_{cat}, by measuring the reaction velocity under conditions of [S]<K_m so that v approaches V_{max}. The rate of enzyme turnover under most physiological conditions in vivo, however, is very different from our laboratory situation.

In vivo, the concentration of substrate is more typically 0.1—1.0K_m. When [S]=K_m we must change our expression for velocity to:

$$v = \frac{k_{cat}}{K_m}[E][S]_f \tag{5.27}$$

(Since [S]<K_m here, the free enzyme concentration is well approximated by the total enzyme concentration [E]; thus we used this term in Equation 5.27 in place of $[E]_f$.) Recalling our definition of K_m, we note that:

$$\frac{k_{cat}}{K_m} = \frac{k_{cat}k_1}{k_{-1} + k_{cat}} \tag{5.28}$$

Thus, under our laboratory conditions, where [S]>K_m, formation of the ES complex is rapid and often is not the rate-limiting step. In vivo, however, where [S]<K_m, the overall reaction may be limited by the diffusional rate of encounter of the free enzyme with substrate, which is defined by k_m. The rate constant for diffusional encounters between molecules like enzymes and substratesis typically in the range of 10^8-$10^9 M^{-1} s^{-1}$. Thus we must keep in mind that the rate-limiting step in catalysis is not always the same in vivo as in vitro. Nevertheless, measurement of k_{cat} (i.e., velocity under saturating substrate concentration) gives us the most consistent means of comparing rates for different enzymatic reactions.

The significance of k_{cat} is that it defines for us the maximal velocity at which an enzymatic reaction can proceed at a fixed concentration of enzyme and infinite availability of substrate. Because k_{cat} relates to the chemical steps subsequent to formation of the ES complex, changes in k_{cat} , brought about by changes in the enzyme (e.g., mutagenesis of specific amino acid residues, or comparison of different enzymes), in solution conditions (e.g., pH, ionic strength, temperature, etc.), or in substrate identity (e.g., structural analogues or isotopically labeled substrates), define perturbations that affect the chemical steps in enzymatic catalysis. In other words, changes in k_{cat} reflect perturbations of the chemical steps subsequent to initial substrate binding. Since k_{cat} reflects multiple chemical steps, it does not provide detailed information on the rates of any of the individual steps subsequent to substrate binding. Instead k_{cat} provides a lower limit on the first-order rate constant of the slowest (i.e., rate-determining) step following substrate binding that leads eventually to product release.

5.5.3 kcat /Km

The catalytic efficiency of an enzyme is best defined by the ratio of the kinetic constants, k_{cat}/K_m. This ratio has units of a second-order rate constant and is generally used to compare the efficiencies of different enzymes to one another. The values of k_{cat}/K_m are also used to compare the utilization of different substrates for a particular enzyme. As we shall see in comparisons of different substrates for an enzyme, the largest differences often are seen in the values of k_{cat}, rather than in K_m. This is because substrate specificity often results from differences in transition state, rather than ground state binding interactions . Hence, the ratio k_{cat}/K_m captures the effects of differing substrate on either kinetic constant and provides a lower limit for the second-order rate constant of productive substrate binding (i.e., substrate binding leading to ES‡ complex formation and eventual product formation); this ratio is therefore considered to be the best measure of substrate specificity.

The ratio k_{cat} $/K_m$ is also used to compare the efficiency with which an enzyme catalyzes a particular reaction in the forward and reverse directions. Enzymatic reactions are in principle reversible, although for many enzymes the reverse reaction is thermodynamically unfavorable. The presence of an enzyme in solution does not alter the equilibrium constant K_{eq} between the free substrate and free product concentrations. Hence, the value of K_{eq} is fixed for specific solution conditions, and this constrains the values of k_{cat}/K_m that can be achieved in the forward (f) and reverse (r) directions. At equilibrium the forward and reverse reactions occur with equal frequency so that:

$$\left(\frac{k_{cat}}{K_m}\right)_f [E][S] = \left(\frac{k_{cat}}{K_m}\right)_r [E][P] \tag{5.29}$$

hence,

$$K_{eq} = \frac{(k_{cat}/K_m)_f}{(k_{cat}/K_m)_r} \tag{5.30}$$

Equation 5.30, known as the Haldane relationship, provides a useful measure of the directionality of an enzymatic reaction under a specific set of solution conditions.

In either direction, the ratio k_{cat}/K_m can be related to the free energy difference between the free reactants (E and S, in the forward direction) and the transition state complex (ES‡). If we normalize the free energy of the reactant state to zero, the free energy difference is defined by:

$$\Delta G_{ES‡} = -RT \ln\left(\frac{k_{cat}}{K_m}\right) + RT \ln\left(\frac{k_B T}{h}\right) \tag{5.31}$$

where k_m is the Boltzmann constant, T is temperature in degrees Kelvin, and h is Planck's constant, so that at a fixed temperature the term $RT \ln(k_B T/h)$ is a constant. This relationship holds because generally attainment of the transition state is the most energetically costly component of the multiple steps contributing to k_{cat}. If we compare different substrates for a single enzyme, or different enzymes or mutants versus a common substrate, we can calculate the difference in transition state energies ($\Delta\Delta G_{ES‡}$) from experimentally measuredvalues of k_{cat}/K_m at constant temperature:

$$\Delta\Delta G_{ES‡} = -RT \ln\left[\frac{(k_{cat}/K_m)^1}{(k_{cat}/K_m)^2}\right] \tag{5.32}$$

where the superscripts 1 and 2 refer to the different substrates or enzymes being compared. By this type of analysis, one can quantitate the thermodynamic contributions of particular structural components to catalysis. For example, suppose one suspected that an active site tyrosine residue was forming a critical hydrogen bond with substrate in the transition state of the enzymatic reaction.

Through the tools of molecular biology, one could replace this tyrosine with a phenylalanine (which would be incapable of forming an H bond) by sitedirected mutagenesis and measure the value of k_{cat}/K_m for both the wild type and the Tyr->Phe mutant. Suppose these values turned out to be 88M^{-1} s^{-1} for the wild-type enzyme and 0.1M^{-1} s^{-1} for the mutant. The ratio of these k_{cat}/K_m would be 880 and, from Equation 5.30, this would correspond to a difference in transition state free energy of 4 kcal/mol, consistent with a strong H-bonding interaction of the tyrosine (of course this does not prove that the exact role of the tyrosine OH group is H-bonding, but the data do prove that this OH group plays a critical role in catalysis). A good example of the use of this approach can be found in the paper by Wilkinson et al. .

5.5.4 Diffusion-Controlled Reactions and Kinetic Perfection

For an enzyme in solution, the rate-determining step in catalysis will be either K_1, the rate of ES formation, or one of the multiple steps contributing to k_{cat}. If k_{cat} is rate limiting, the catalytic events that occur after substrate binding are slower that the rate of formation of the ES complex. If, however, k_1 is rate limiting, the enzyme turns over essentially instantaneously once the ES complex has formed. In either case we see that the fastest rate of catalysis for an enzyme in solution is limited by the rate of diffusion of molecules in the solution. Some enzymes, such as carbonic anhydrase, display k_{cat}/K values of 10^8-10^9 M^{-1} s^{-1}, which is at the diffusion limit. Such enzymes are said to have achieved *kinetic perfection*, because they convert substrate to product as fast as the substrate is delivered to the active site of the enzyme!

The diffusion limit would seem to set an upper limit on the value of k_{cat}/K_m that an enzyme can achieve. This is true for most enzymes in solution. However, some enzyme systems have overcome this limit by compartmentaliz- ing themselves and their substrates within close proximity in subcellular locals where three-dimensional diffusion no longer comes into play. This can be accomplished by assembling enzymes and substrates into organized systems such as multienzyme complexes or cellular membranes. Two examples are presented.

We first consider the respiratory electron transfer system of the inner mitochondrial membrane. Here enzymes in a cascade are localized in close proximity to one another within the membrane bilayer. The product of one enzyme is the substrate for the next in the cascade. Because of the proximity of the enzymes in the membrane, the product leaves the active site of one enzyme and is presented to the active site of the next enzyme without the need for diffusion through solution.

The second example comes from the de novo synthetic pathway for pyrimidines. The first three steps in the synthesis of uridine monophosphate are performed by a supercomplex of three enzymes that are noncovalently associated as a multiprotein complex. This supercomplex, referred to as CAD, comprises the enzymes carbamyl phosphate synthase, aspartate transcarbamylase, and dihydroorotase. Because the

active sites of the three enzymes are compartmentalized inside the supercomplex, the product of the first enzyme is immediately in proximity to the active site of the second enzyme, and so on. In this way, the supercomplex can overcome the diffusion barrier to rapid catalysis.

5.6 EXPERIMENTAL MEASUREMENT OF KCAT AND KM

5.6.1 Graphical Determinations from Untransformed Data

The kinetic constants V_{max} and K_m are determined graphically with initial velocity measurements obtained at varying substrate concentrations. The graphical methods are best illustrated by working through examples with some numerical data. The quality of the estimates of V_{max} and K_m depend on covering a substrate concentration range that spans a significant portion of the binding isotherm. Experimentally, a convenient method for choosing substrate concentrations is to first make a stock solution of substrate at the highest concentration that is experimentally reasonable. Then, twofold serial dilutions can be made from this stock to produce a range of lower substrate concentrations. For example, let us say that the highest concentration of substrate to be used in an enzymatic reaction is 250 μM. We could make a 2.5mM stock solution of the substrate that would be diluted 10-fold into the final assay reaction mixture (i.e., to give a final concentration of 250 μM). We could then take a portion of this stock solution and dilute it with an equal volume of buffer to yield a 1.25mM solution, which upon dilution into the assay reaction mixture would give a final substrate concentration of 125 μM. A portion of this solution could also be diluted in half with buffer, and so on, to yield a series of solutions of diminishing substrate concentrations. The final substrate concentrations presented in Table 5.1 illustrate the use of such a twofold *serial dilution* scheme. Let us suppose that we are studying a simple enzymatic reaction for which the true values of V_{max} and K_m are 100 μM/s and 12 μM, respectively. In Table 5.1 we have listed experimental values for the initial velocity v at each of the substrate concentrations used. In generating this table, some random error has been added to each of the velocity values to better simulate real experimental conditions. The largest percent errors in this table occur at the lowest substrate concentrations, where in real experiments one encounters the greatest difficulty in obtaining accurate velocity measurements.

Table 5.1 Initial velocity (with random error added) as a function of substrate concentration for a model enzymatic reaction

$[S](\mu M)^a$	$v(\mu M$ product formed $s^{-1})$	$1/v(\mu M^{-1} \cdot s)$	$1/[S](\mu M^{-1})$
0.98	10	0.100	1.024
1.95	12	0.083	0.512
3.91	28	0.036	0.256
7.81	40	0.025	0.128
15.63	55	0.018	0.064
31.25	75	0.013	0.032
62.50	85	0.012	0.016
125.00	90	0.011	0.008
250.00	97	0.010	0.004

*Substrate concentrations reflect a twofold serial dilution starting with an initial solution that provides 250 μM substrate to the final assay reaction mixture.

The first and most straightforward way of graphing the data is as a direct plot of velocity as a function of [S]; we shall refer to such a plot as a Michaelis–Menten plot. Figure 5.5 is a Michaelis–Menten plot for the data in Table 5.1; the line drawn through the data was generated by a nonlinear least-squares fit of the data to Equation 5.24. With modern computer graphics programs, the reader has a wide choice of options for performing such curve fitting; some programs that are particularly well suited for enzyme studies are listed in Appendix II. The plots in this book, for example, were generated with the program Kaleidagraph (from Abelbeck Software), which contains a built-in iterative method for performing nonlinear curve fitting to user-generated equations. For the data in Figure 5.5 both V_{max} and K_m were set as unknowns that were simultaneously solved for by the curve-fitting routine. The estimates of V_{max} and K_m determined in this way were 100.36 μM/s and 11.63 μM, respectively, in excellent agreement with the true values of these constants.

Such direct fits of the untransformed data provide the most reliable estimates of both kinetic constants. With the widespread availability of computer curve-fitting programs, what limitations are there on our ability to estimate V_{max} and K_m from experimental data points represents the nonlinear least-squares best fit to Equation 5.24.

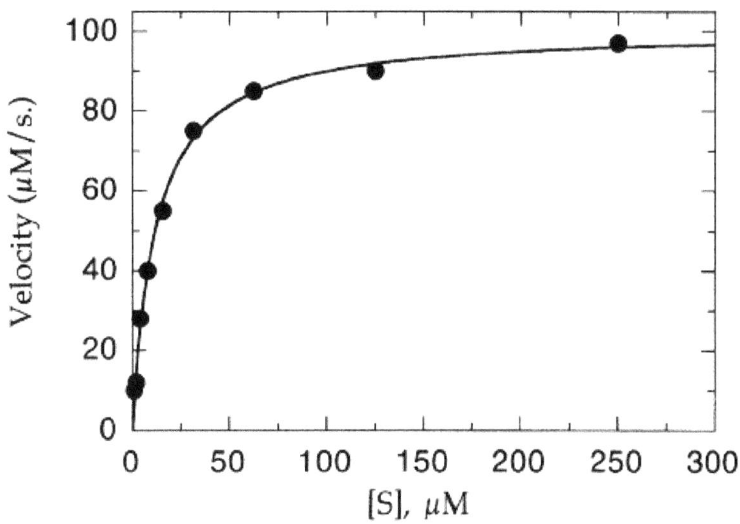

Figure 5.5 Michaelis–Menten plot for the velocity data in Table 5.1. The solid line through the data? As mentioned above, the accuracy of such estimates will depend on the range of substrate concentrations over which the initial velocity has been determined. If measurements are made only at low substrate concentrations, the data will appear to be first-ordered (i.e., *v* will appear to be a linear function of [S]). This is illustrated in Figure 5.6A for the data in Table 5.1 between substrate concentrations of 0.98 and 3.91 μM (i.e., <0.33K_m). In this concentration range, the enzyme active sites never reach saturation, and graphically, both V_{max} and K_m appear to be infinite . On the other hand, Figure 5.6B illustrates what happens when measurements are made at very high substrate concentrations only; here the data for substrate concentrations above 60 μM are considered (i.e., [S]>5K_m). In this saturating substrate concentration range, the velocity appears to be almost independent of substrate concentration. While a rough estimate of V_{max} might be obtained from these data (although the reader should note that the true V_{max} is only reached at infinite substrate concentration; hence any experimentally measured velocity at high [S] may approach, but never fully reach V_{max}), there is no way to determine the K_m value here.

The plots in Figure 5.6 emphasize the need for exploring a broad range of substrate concentrations to accurately determine the kinetic constants for the enzyme of interest. Again, there may be practical limits on the range of substrate concentrations over which such measurements can be performed. In Chapter 4 we suggested that to best characterize a ligand binding isotherm, it is necessary to cover a ligand concentration range that resulted in 20–80% receptor saturation. Likewise, in determining the steady state kinetic constants for an enzymatic system, it is best to at least cover substrate concentrations that yield velocities of 20—80% of V_{max}; this corresponds to [S] of 0.25—5.0K_m.

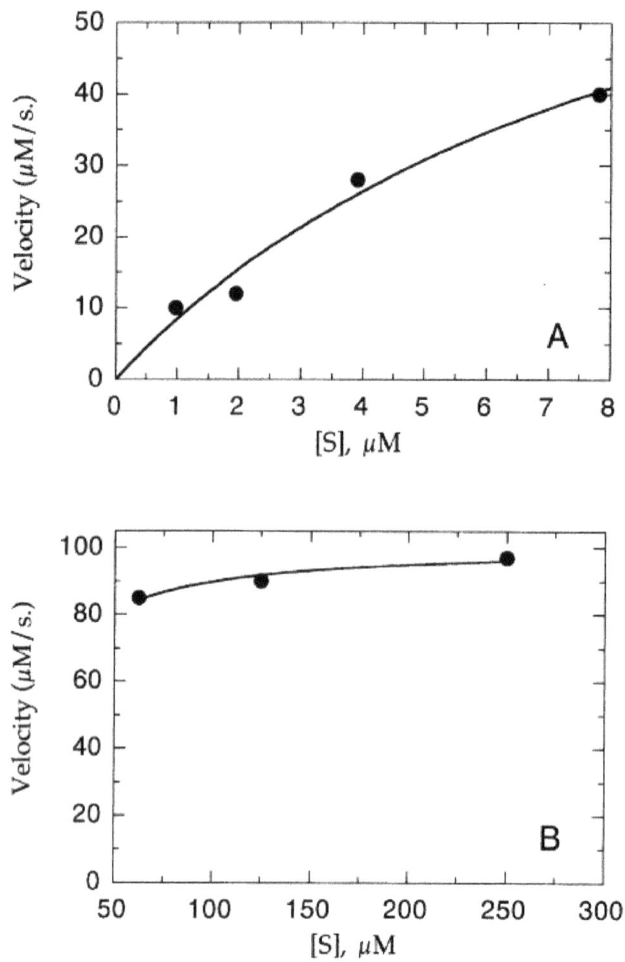

Figure 5.6 Michaelis—Menten plots for restricted data from Table 5.1. (A) The range of [S] values is inappropriately low (<0.33Km), hence Km and Vmax appear to be infinite. (B) The range of [S] values is inappropriately high, with the result that every data point represents nearsaturating conditions; one may be able to approximate Vmax, but Km cannot be determined.

Since the kinetic constants are unknowns prior to these experiments, it is common to perform initial experiments with a limited number of data points that span as broad a range of substrate concentrations as possible (at least a 100-fold substrate concentration range) to obtain a rough estimate of K_m and V_{max}. Improved estimates can then be obtained by narrowing the substrate concentration range between 0.25 and $5.0K_m$ and obtaining a larger number of data points within this range. Table 5.1 illustrates an ideal situation for determining V_{max} and K_m. The data span a 250-fold range of substrate concentrations that cover the range from 0.08 to $20.8K_m$. Hence, use of such a twofold serial dilution setup, starting with the highest substrate concentration that is feasible, is highly recommended.

Alternatively, one could perform a limited number of experiments by using a fivefold serial dilution setup starting at our maximum substrate concentration of 250 µM. With only five experiments we would then cover the substrate concentrations 250, 50, 10, 2, and 0.4 µM. Data from such a hypothetical experiment are shown in Figure 5.7A and would yield an estimate of K_m of about 10 µM. With this initial estimate in hand, one might then chose to expand the number of data points within the narrower range of 0.25—$5.0K_m$ to obtain better estimates of the kinetic constants. Figure 5.7B, for example, illustrates the type of data one might obtain from velocity measurements at substrate concentrations of 2.5, 5, 10, 15, 20, 25, 30, 35, 40, 45, and 50 µM. From this second set of measurements, values of 102 M/s and 12 µM for V_{max} and K_m, respectively, would be obtained.

5.6.2 Lineweaver–Burk Plots of Enzyme Kinetics

The widespread availability of user-friendly nonlinear curve-fitting programs is a relatively recent development. In the past, determination of the kinetic constants for an enzyme from the untransformed data was not so routine. To facilitate work in this area, scientists searched for means of transforming the data to produce linear plots from which the kinetic constants could be determined simply with graph paper and a straightedge. While today many of us have nonlinear curve-fitting programs at our disposal (and this is the preferred means of determining the values of V_{max} and K_m), there is still considerable value in linearized plots of enzyme kinetic data. As we shall see in subsequent chapters, these plots are extremely useful in diagnosing the mechanistic details of multisubstrate enzymes and for determining the mode of interaction between an enzyme and an inhibitor.

The most commonly used method for linearizing enzyme kinetic data is that of Lineweaver and Burk . We start with the same steady state assumption described earlier. Applying some simple algebra, we can rewrite Equation 5.24 in the following form:

$$v = V_{max}\left(\frac{1}{1+\frac{K_m}{[S]}}\right) \quad (5.33)$$

Now we simply take the reciprocal of this equation and rearrange to obtain:

$$\frac{1}{v} = \left(\frac{K_m}{V_{max}}\frac{1}{[S]}\right) + \frac{1}{V_{max}} \quad (5.34)$$

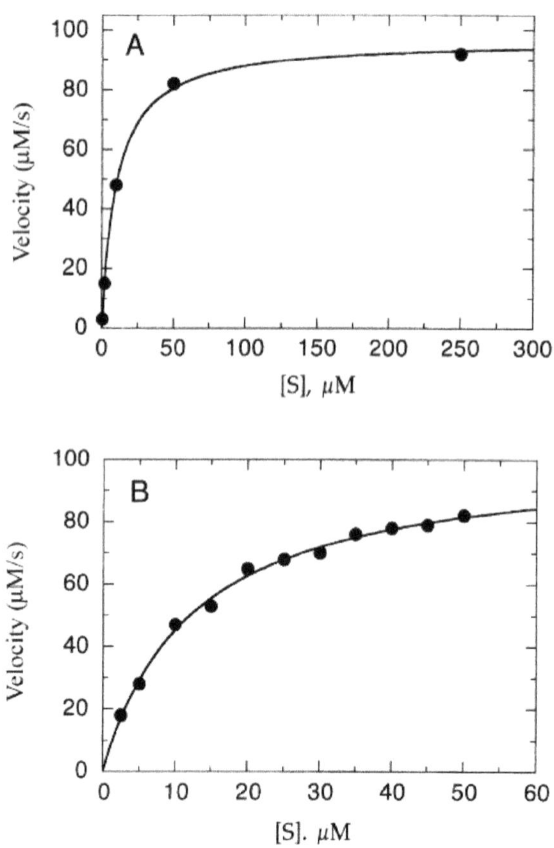

Figure 5.7 Experimental strategy for estimating Km and Vmax. (A) A limited data set is collected over a broad range of [S] to get a rough estimate of the kinetic constants. (B) Once a rough estimate of Km has been determined, a second set of experiments is performed with more data within the range of 0.25–5.0Km to obtain more precise estimates of the kinetic constants.

Comparing Equation 5.34 with the standard equation for a straight line, we have

$$y = mx + b \tag{5.35}$$

where m is the slope and b is the y intercept. We see that Equation 5.34 is an equation for a straight line with slope of K_m/V_{max} and y intercept of 1/Vmax

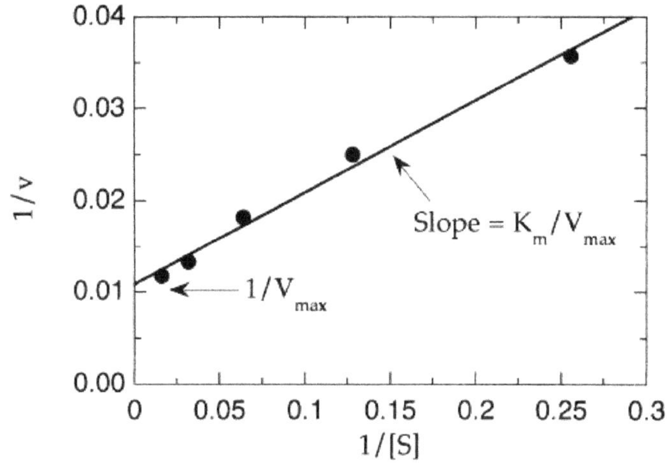

Figure 5.8 Lineweaver–Burk double-reciprocal plot for selected data from Table 5.1 within the range of [S]=0.25–5.0Km.

Thus if the reciprocal of initial velocity is plotted as a function of the reciprocal of [S], we would expect from Equation 5.34 to obtain a linear plot. For the same reasons described earlier for untransformed data, these plots work best when the substrate concentration covers the range of 0.25–5.0K_m. Within this range, good linearity is observed, as illustrated in Figure 5.8 for the data between [S]=3.91 and [S]=62.50 μM in Table 5.1. A plot like that in Figure 5.8 is known as a Lineweaver–Burk plot.

The kinetic constants K_m and V_{max} can be determined from the slope and intercept values of the linear fit of the data in a Lineweaver–Burk plot. Since the x axis is reciprocal substrate concentration, the value of x=0 (i.e., 1/[S]=0) corresponds to [S]=infinity. Hence, the extrapolated value of the y intercept corresponds to the reciprocal of V_{max}. The value of K_m can be determined from a Lineweaver–Burk plot in two ways. First we note from Equation 5.34 that the slope is equal to K_m divided by V_{max}. If we therefore divide the slope of our best fit line by the y-intercept value (i.e., by $1/V_{max}$), the product will be equal to K_m. Alternatively, we could extrapolate our linear fit to the point of intersecting the x axis. This x intercept is equal to $-1/K_m$; thus we could determine K_m from the absolute value of the reciprocal of the x intercept of our plot.

We have noted several times that the preferred way to determine K_m and V_{max} values is from nonlinear fitting of untransformed data to the Michaelis— Menten equation. Figure 5.8 illustrates why we have stressed this point. In real experimental data, small errors in the measured values of v are amplified by the mathematical transformation of taking the reciprocal. The greatest percent error is likely to be associated with velocity values at low substrate concentration. Unfortunately, in the reciprocal plot, the lowest values of [S] correspond to the highest values of 1/[S], and because of the details of linear regression, these data points are weighted more heavily in the analysis. Hence the experimental error is amplified and unevenly weighted in this analysis, resulting in poor estimates of the kinetic constants even when the experimental error is relatively small. To illustrate this, let us compare the estimates of V_{max} and K_m obtained for the data in Table 5.1 by various graphical methods; this is summarized in Table 5.2. The true values of V_{max} and K_m for the hypothetical data in Table 5.1 were 100 μM/s and 12 μM, respectively. The fitting of the untransformed data to the Michaelis—Menten equation provided estimates of 100.36 and 11.63 for the two kinetic constants, with deviations from the true values of only 0.36 and 3.08%, respectively. The linear fitting of the data in Figure 5.8, on the other hand, yields estimates of V_{max} and K_m of 91.84 and 9.17, with deviations from the true values of 8.16 and 23.58%, respectively. The errors are even greater when the double-reciprocal plots are used for the full data set in Table 5.1, as illustrated in Figure 5.9 and Table 5.2. Here the inclusion of the low substrate data values (<[S]=3.91 μM) are very heavily weighted in the linear regression and further limit the precision of the kinetic constant estimates. The values of V_{max} and K_m derived from this fitting are 79.28 and 7.57, representing deviations from the true values of 20.72 and 36.92%, respectively.

Table 5.2 Estimates of the kinetic constants Vmax and Km from various graphical treatments of the data from Table 5.1

Graphical Method	K_m (μM)	Deviation from True K_m (%)	V_{max} (μM/s)	Deviation from True V_{max} (%)
True values	12.00		100.00	
Michaelis–Menten	11.63	3.08	100.36	0.36
Lineweaver–Burk (full data set)	7.57	36.92	79.28	20.72
Lineweaver–Burk ([S] = 0.25–5.0K_m only)	9.17	23.58	91.84	8.16
Eadie–Hofstee	9.66	19.50	94.45	5.55
Hanes–Wolff	11.84	1.33	100.97	0.97
Eisenthal–Cornish-Bowden	11.53	3.92	100.64	0.64

The foregoing example, should convince the reader of the limitations of using linear transformations of the primary data for determining the values of the kinetic constants. Nevertheless, the Lineweaver–Burk plots are still commonly used by many researchers and, as we shall see in later chapters, are valuable tools for certain purposes. In these situations , we make the following recommendation. Rather than using linear regression to fit the reciprocal data in Lineweaver–Burk plots, one should determine the values of V_{max} and K_m by nonlinear regression analysis of the untransformed data fit to the Michaelis–Menten equation. These values are then inserted as constants into Equation 5.34 to create a line through the reciprocal data on the Lineweaver–Burk plot. The line drawn by this method may not appear to fit the reciprocal data as well as a linear regression fit, but it will be a much more accurate reflection of the kinetic behavior of the enzyme.

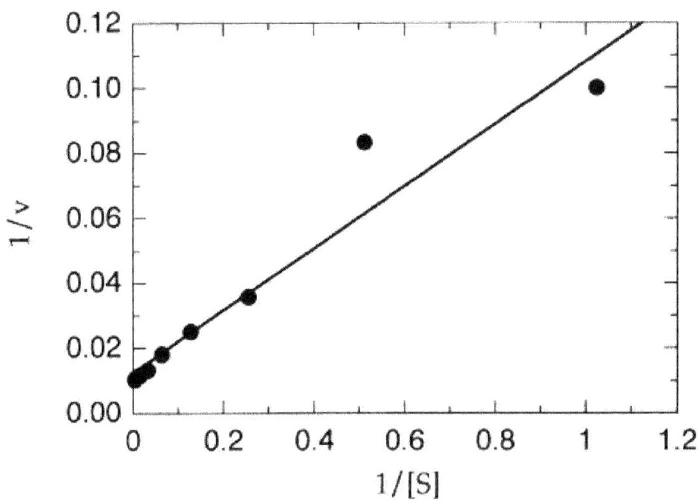

Figure 5.9 Lineweaver–Burk double-reciprocal plot for the full data set from Table 5.1. Note the strong influence of the data points at low [S] (high 1/[S] values) on the best fit line from linear regression.

If one is to ultimately present experimental data in the form of a doublereciprocal plot, it is desirable to chose substrate concentrations that will be evenly spaced along a reciprocal *x* axis (i.e., 1/[S]). This is easily accomplished experimentally as follows. One picks a maximum value of [S] ($[S_{max}]$) to work with and makes a stock solution of substrate that will give this final concentration after dilution into the assay reaction mixture. Additional initial velocity measurements are then made by adding the same final volume to the enzyme reaction mixture from stock substrate solutions made by diluting the original stock solution by 1:2, 1:3, 1:4, 1:5, and so on. In this way, the data points will fall along the 1/[S] axis at intervals of 1, 2, 3, 4, 5, . . . , units.

For example, let us say that we have decided to work with a maximum substrate concentration of 60 μM in our enzymatic reaction. If we prepare a 600 μM stock solution of substrate for this data point, we might dilute it 1:10 into our assay reaction mixture to obtain the desired final substrate concentration. If, for example, our total reaction volume were 1.0 mL, we could start our reaction by mixing 100 μL of substrate stock, with 900 μL of the other components of our reaction system (enzyme, buffer, cofactors, etc.). Table 5.3 summarizes the additional stock solutions that would be needed to prepare final substrate concentrations evenly spaced along a 1/[S] axis.

Table 5.3 Setup for an experimental determination of enzyme kinetics using a Linewever--Burk plot

Stock [S] (μM)	Final [S] in Reaction Mixture (μM)	1/[S] (μM^{-1})
600	60.0	0.017
300	30.0	0.033
200	20.0	0.050
150	15.0	0.067
120	12.0	0.083
100	10.0	0.100
86	8.6	0.116
75	7.5	0.133
67	6.7	0.149
60	6.0	0.167
55	5.5	0.182
50	5.0	0.200

5.7 Other Linear Transformations of Enzyme Kinetic Data

Despite the errors associated with this method, the Lineweaver—Burk double reciprocal plot has become the most popular means of graphically representing enzyme kinetic data. There are, however, a variety of other linearizing transformations.

Again, the use of these transformation methods is no longer necessary because most researchers have access to computer-based nonlinear curve-fitting methods, and the direct fitting of untransformed data by these methods is highly recommended. For the sake of historic persepective, however, we shall describe three other popular graphical methods for presenting enzyme kinetic data: Eadie—Hofstee, Hanes—Wolff, and Eisenthal—Cornish- Bowden direct plots. These linear transformation methods, which are here applied to enzyme kinetic data, are identical to the Wolff transformations for receptor—ligand binding data.

5.7.1 Eadie–Hofstee Plots

If we multiply both sides of Equation 5.24 by K_m +[S], we obtain:

$$v(K_m + [S]) = V_{max}[S] \tag{5.36}$$

If we now divide both sides by [S] and rearrange, we obtain:

$$v = V_{max} - K_m \left(\frac{v}{[S]}\right) \tag{5.37}$$

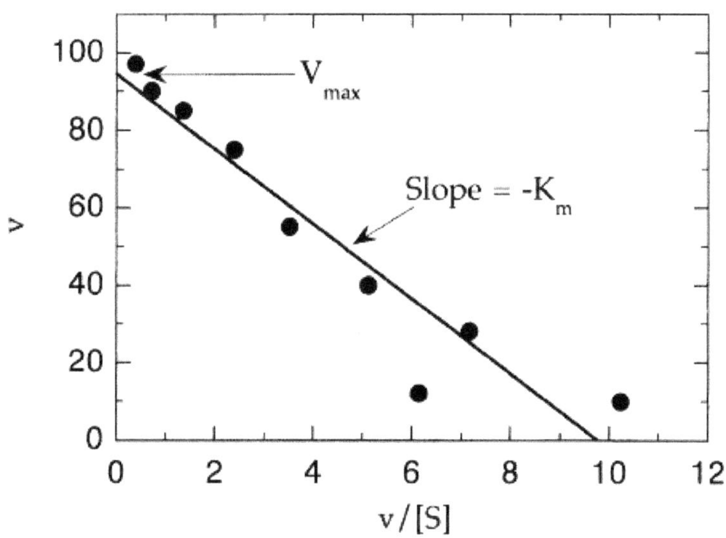

Figure 5.10 Eadie—Hofstee plot of enzyme kinetic data. Data taken from Table 5.1.

Hence, if we plot *v* as a function of *v*/[S], Equation 5.37 would predict a straight line relationship with slope of -K_m and *y* intercept of V_{max}. Such a plot, referred to as an Eadie—Hofstee plot, is illustrated in Figure 5.10.

5.7.2 Hanes–Wolff Plots

If one multiplies both sides of the Lineweaver—Burk Equation (Equation 5.34) by [S], one obtains:

$$\frac{[S]}{v} = [S]\left(\frac{1}{V_{max}}\right) + \frac{K_m}{V_{max}} \tag{5.38}$$

This treatment also leads to linear plots when [S]/v is plotted as a function of [S]. Figure 5.11 illustrates such a plot, which is known as a Hanes–Wolff plot. In this plot the slope is $1/V_{max}$, the y intercept is K_m/V_{max}, and the x intercept is $-K_m$.

5.7.3 Eisenthal–Cornish-Bowden Direct Plots

In our final method, pairs of v, [S] data (as in Table 5.1) are plotted as follows: values of v along the y axis and the negative values of [S] along the x axis . For each pair, one then draws a straight line connecting the points on the two axes and extrapolates these lines past their point of intersection (Figure 5.12). When a horizontal line is drawn from the point of intersection of these line to the y axis, the value at which this horizontal line crosses the y axis is equal to V_{max}. Similarly, when a vertical line is dropped from the point of intersection to the x axis, the value at which this vertical line crosses the x axis defines K_m. Plots like Figure 5.12, are referred to as Eisenthal–Cornish-Bowden direct plots and are considered to give the best estimates of K_m and V_{max} of any of the linear transformation methods. Hence they are highly recommended when it is desired to determine these kinetic parameters but nonlinear curve fitting to Equation 5.24 is not feasible.

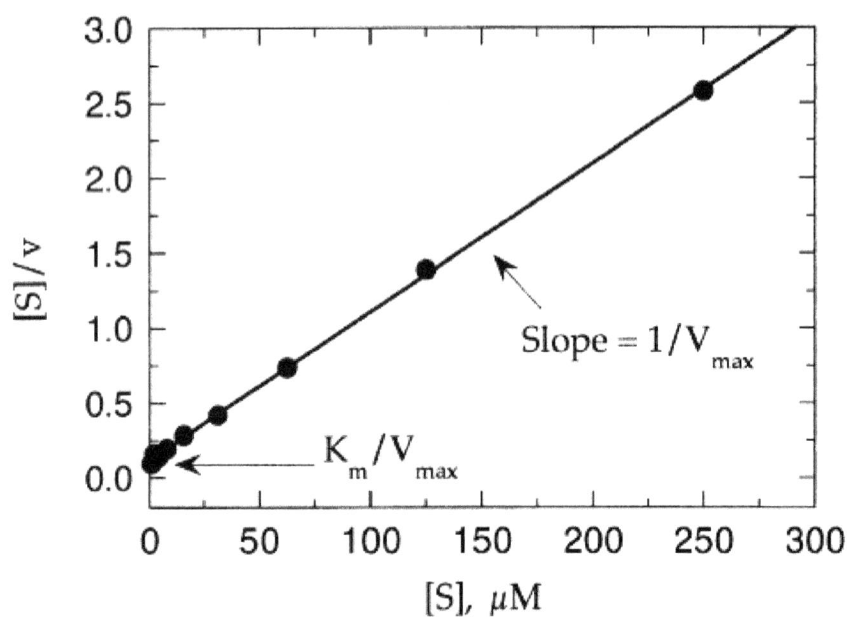

Figure 5.11 Hanes–Wolff plot of enzyme kinetic data. Data taken from Table 5.1.

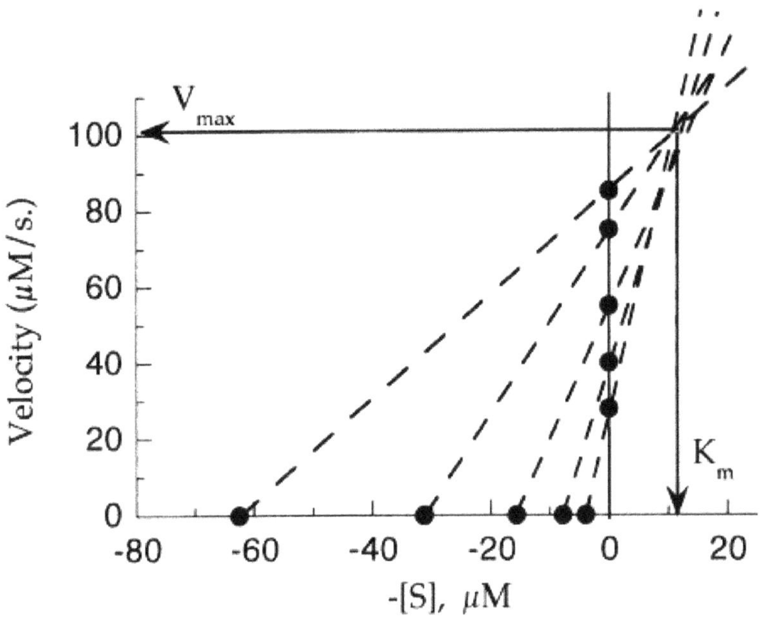

Figure 5.12 Eisenthal—Cornish-Bowden direct plot of enzyme kinetic data. Selected data taken from Table 5.1.

5.8 MEASUREMENTS AT LOW SUBSTRATE CONCENTRATIONS

In some instances the concentration range of substrates suitable for experimental measurements is severely limited because of poor solubility or some physicochemical property of the substrate that interferes with the measurements above a critical concentration. If one is limited to measurements in which the substrate concentration is much less than the K_m, the reaction will follow pseudo-first-order kinetics, and it may be difficult to find a time window over which the reaction velocity can be approximated by a linear function. Even if quasi-linear progress curves can be obtained, a plot of initial velocity as a function of [S] cannot be used to determine the individual kinetic constants k_{cat} and K_m, since the substrate concentration range that is experimentally attainable is far below saturation (as in Figure 5.6A). In such situations one can still derive an estimate of k_{cat}/K_m by fitting the reaction progress curve to a first-order equation at some fixed substrate concentration.

Suppose that we were to follow the loss of substrate as a function of time under first-order conditions (i.e., where $[S]<K_m$). The progress curve could be fit by the following equation:

$$[S] = [S_0]e^{-kt} \tag{5.39}$$

where [S] is the substrate concentration remaining after time t, $[S_0]$ is thestarting concentration of substrate, and k is the observed first-order rate constant. When $[S]<K_m$, the [S] term can be ignored in the denominator of Equation 5.24. Combining this with our definition of V_{max} from Equation 5.10 we obtain:

$$-\frac{d[S]}{dt} = \frac{k_{cat}}{K_m}[E][S] \tag{5.40}$$

Rearranging Equation 5.40 and integrating, we obtain:

$$[S] = [S_0]\exp\left(\frac{k_{cat}}{K_m}[E]t\right) \tag{5.41}$$

Comparing Equation 5.41 with Equation 5.39, we see that:

$$k = \frac{k_{cat}}{K_m}[E] \tag{5.42}$$

Thus if the concentration of enzyme used in the reaction is known, an estimate of k_{cat}/K_m can be obtained from the measured first-order rate constant of the reaction progress curve when $[S]>K_m$.

5.9 DEVIATIONS FROM HYPERBOLIC KINETICS

In most cases enzyme kinetic measurements fit remarkably well to the Henri–Michaelis–Menten behavior discussed in this chapter. However, occasional deviations from the hyperbolic dependence of velocity on substrate concentration are seen. Such anomalies occur for several reasons. Some physical methods of measuring velocity, such as optical spectroscopies, can lead to experimental artifacts that have the appearance of deviations from the expected behavior.

Nonhyperbolic behavior can also be caused by the presence of certain types of inhibitor as well. In the most often encountered case, substrate inhibition, a second molecule of substrate can bind to the ES complex to form an inactive ternary complex, SES. Because formation of the ES complex must precede formation of the inhibitory ternary complex, substrate inhibition is usually realized only at high substrate concentrations, and it is detected as a lower than expected value for the measured velocity at these high substrate concentrations.

Figure 5.13 illustrates the type of behavior one might see for an enzyme that exhibits substrate inhibition. At low substrate concentrations, the kinetics follow simple Michaelis—Menten behavior. Above a critical substrate concentration, however, the data deviate significantly from the expected behavior. The binding of the second, inhibitory, molecule of substrate can be accounted for by the following equation:

$$v = \frac{V_{max}[S]}{K_m + [S]\left(1 + \frac{[S]}{K_i}\right)} \tag{5.43}$$

or dividing the top and bottom of the right-hand side of Equation 5.43 by [S], we obtain:

$$v = \frac{V_{max}}{1 + \frac{K_m}{[S]} + \frac{[S]}{K_i}} \tag{5.44}$$

where the term K_i in Equations 5.43 and 5.44 represents the dissociation constant for the inhibitory SES ternary complex. Inhibition effects at very high substrate concentrations also can be readily detected as nonlinearity in the Lineweaver—Burk plots of the data. Here one observes a sudden and dramatic curving up of the data near the *y*-axis intercept.

Another cause of nonhyperbolic kinetics is the presence of more than one enzyme acting on the same substrate . Many enzyme studies are performed with only partially purified enzymes, and many clinical diagnostic tests that rely on measuring enzyme activities are performed on crude samples (of blood, tissue homogenates, etc.). When the substrate for the reaction is unique to the enzyme of interest, these crude samples can be used with good results. If, however, the sample contains more than one enzyme that can act on the substrate, deviations from the expected kinetic results occur. Suppose that our sample contains two enzymes; both can convert the substrate to product, but they display different kinetic constants. Suppose further that for one of the enzymes $V_{max}V_1$ and K_mK_1, and for the second enzyme $V_{max}=V_2$ and K_mK_2 . The velocity of the overall mixture then is given by:

$$v = \frac{V_1[S]}{K_1 + [S]} + \frac{V_2[S]}{K_2 + [S]} \tag{5.45}$$

This can be rearranged to give the following expression :

$$v = \frac{(V_1K_2 + V_2K_1)[S] + (V_1 + V_2)[S]^2}{K_1K_2 + (K_1 + K_2)[S] + [S]^2} \tag{5.46}$$

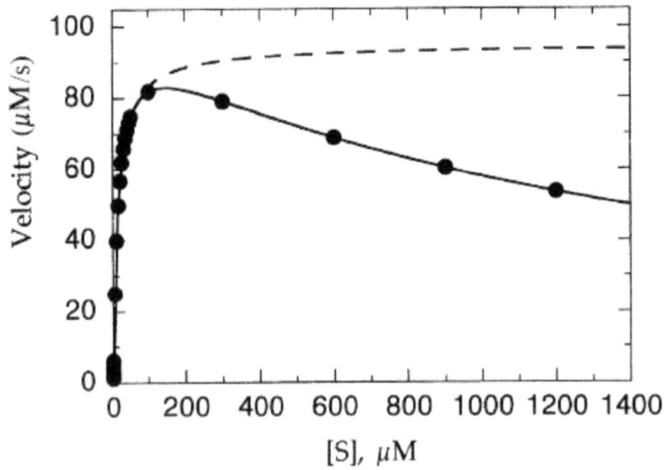

Figure 5.13 Michaelis–Menten plot for an enzyme reaction displaying substrate inhibition at high substrate concentrations: dashed line, best fit of the data at low substrate concentrations to Equation 5.24; solid line, fit of all the data to Equation 5.44. The constant Ki (Equation 5.44) will be described further in subsequent chapters.

Equation 5.46 is a polynomial expression, which yields behavior very different from the rectangular hyperbolic behavior we expect; this is illustrated in Figure 5.14.

One last example of deviation from hyperbolic kinetics is that of enzymes displaying cooperativity of substrate binding. In the derivation of Equation 5.24 we assumed that the active sites of the enzyme molecules behave independently of one another. As we saw in Chapter 3 and 4, sometimes proteins occur as multimeric assemblies of subunits. Some enzymes occur as homomultimers, each subunit containing a separate active site. It is possible that the binding of a substrate molecule at one of these active sites could influence the affinity of the other active sites in the multisubunit assembly . This effect is known as *cooperativity*. It is said to be *positive* when the binding of a substrate molecule to one active site increases the affinity for substrate of the other active sites. On the other hand, when the binding of substrate to one active site lowers the affinity of the other active sites for the substrate, the effect is called *negative cooperativity*. The number of potential substrate binding sites on the enzyme and the degree of cooperativity among them can be quantified by the Hill coefficient, *h*. The influence of cooperativity on the measured values of velocity can be easily taken into account by modifying Equation 5.24 as follows:

$$v = \frac{V_{max}[S]^h}{K' + [S]^h} \tag{5.47}$$

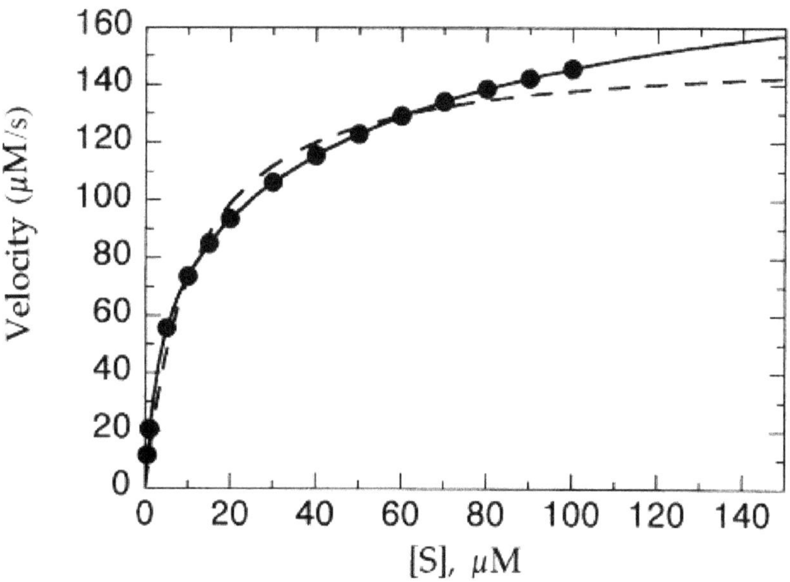

Figure 5.14 Effects of multiple enzymes acting on the same substrate. The dashed line represents the fit of the data to Equation 5.24 for a single enzyme, while the solid line represents the fit to Equation 5.46 for two enzymes acting on the same substrate with $V_1 = 120\ \mu M/s$, $V_2 = 75\ \mu M/s$, $K_1 = 65\ \mu M$, and $K_2 = 3\ \mu M$.

where K' is related to K_m but also contains terms related to the effect of substrate occupancy at one site on the substrate affinity of other sites . Figure 5.15 illustrates how positive cooperativity can affect the Michaelis–Menten and Lineweaver–Burk plots of an enzyme reaction. The velocity data for cooperative enzymes can be presented in a linear form by use of Equation 5.48:

$$\log\left(\frac{v}{V_{max} - v}\right) = h\log[S] - \log(K') \tag{5.48}$$

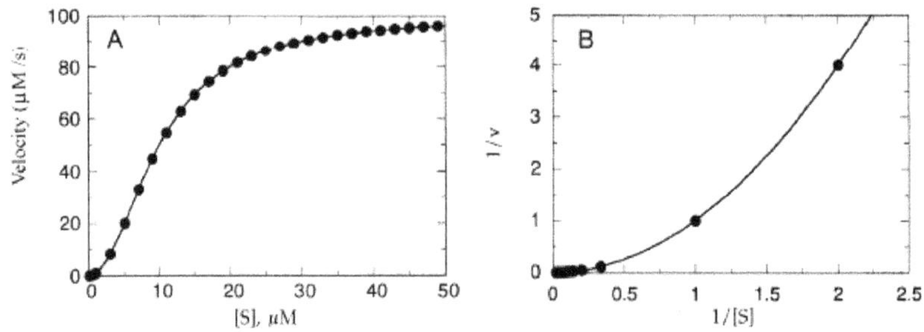

Figure 5.15 Effects of positive cooperativity on the kinetics of an enzyme-catalyzed reaction: (A) data graphed as a Michaelis–Menten (i.e., direct) plot and (B) data from (A) replotted as a Lineweaver–Burk double-reciprocal plot.

Thus, a plot of $\log(v/(V_{max} - v))$ as a function of log[S] should yield a straight line with slope of h and a y intercept of -log(K'), as illustrated in Figure 5.16.

The utility of such plots is limited, however, by the need to know V_{max} a priori and because the linear relationship described by Equation 5.48 holds over only a limited range of substrate concentrations (in the region of [S]=K'). Hence, whenever possible, it is best to determine V_{max}, h, and K_m for cooperative enzymes from direct nonlinear curve fits to Equation 5.47.

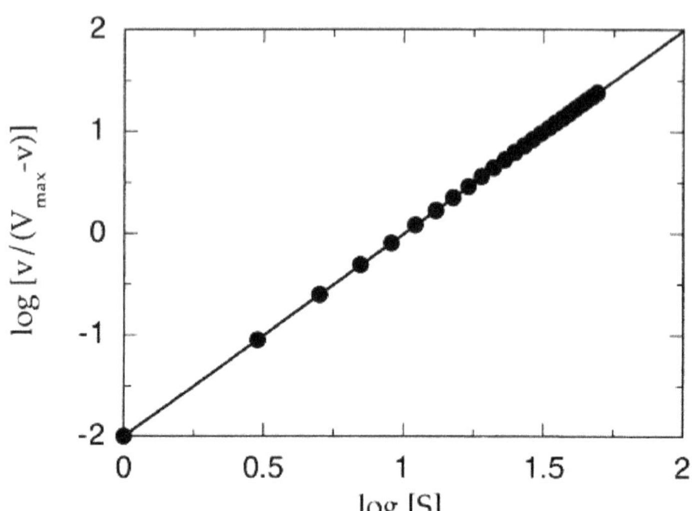

Figure 5.16 Hill plots for the data from Figure 5.15: log[v/(Vmax -v)] is plotted as a function of log[S]. The slope of the best fit line provides an estimated of the Hill coefficient h, and the y intercept provides an estimate of -log(K").

5.10 TRANSIENT STATE KINETIC MEASUREMENTS

Much of the enzymology literature describes studies of the steady state kinetics of enzymatic reactions, and a great deal of biochemical insight can be derived from such studies. Steady state kinetics, however, does have some limitations. The steady state kinetic constants k_{cat} and K_m are complex functions that combined rate constants from multiple steps in the overall enzymatic reaction. Hence, these constants do not provide rate information on any individual steps in the reaction pathway. For example, let us again consider a simple-single substrate enzymatic reaction (as discussed above), but this time let us define the individual rate constants for each step:

$$E + S \underset{k_{-1}}{\overset{k_1}{\rightleftharpoons}} ES \underset{k_{-2}}{\overset{k_2}{\rightleftharpoons}} EX \underset{k_{-3}}{\overset{k_3}{\rightleftharpoons}} EP \underset{k_{-4}}{\overset{k_4}{\rightleftharpoons}} E + P$$

In this scheme, EX represents some transient intermediate in the enzymatic reaction. This could be a distinct chemical species (e.g., an acyl—enzyme intermediate as in peptide hydrolysis by serine proteases; see Chapter 6) or a kinetically significant conformational state that is formed by a change in structure of the ES complex prior to the chemical steps of catalysis. A steady state kinetic study would define the overall reaction in terms of k_{cat} and K_m but would not provide much information on the rates and nature of the individual steps in the reaction. Steady state kinetics also limits the mechanistic detail that one can derive.. These studies do not, however, give information on the specific reactions of the various enzyme species involved in catalysis, nor can they identify the rate-limiting step. The steady state kinetic constant K_m is often confused with the dissociation constant for the ES complex (K_s), and k_{cat} is often mistakenly thought of as a specific rate constantfor a rate-limiting step in catalysis (i.e., k_2); we have already discussed the correct interpretation of K_m and k_{cat}.

To overcome some of these limitations, researchers turn to rapid kinetic methods that allow them to make measurements on a time scale (i.e., milliseconds) consistent with the approach to steady state (pre—steady state kinetics) and to measure the kinetics of transient species that occur after initial substrate binding. These methods are collectively referred to as transient state kinetics, and their application to enzymatic systems provides much richer kinetic detail than simple steady state measurement .

Specialized apparatus must be used to measure kinetic events on a millisecond time scale. A variety of instruments have been designed for this purpose , but the two most commonly used methods for measuring transient kinetics are stopped-flow and rapid reaction quenching. In a stopped-flow experiment, the researcher is measuring the formation of a transient species by detecting a unique optical signal (typically absorbance or fluorescence). Figure 5.17 illustrates a typical design for a stopped-flow apparatus. The instrument consists of two syringes, one holding a enzyme solution and the other holding a substrate solution. Both syringes are attached to a common drive bar that compresses the plungers of both syringes at a steady and common rate, forcing the solutions from each syringe to mix in the mixing chamber and flow through the detection tube. A third syringe is located at the end of the detection tube. As liquid is forced into this third syringe, its plunger is pushed back until it is stopped by contact with a stopping bar. The stopping bar has attached to it a microswitch, which triggers the controlling computer to initiate observation of the optical signal from the solution trapped in the detection tube. Measurements of the optical signal are then made over time as the solution ages in the detection tube.

In a rapid quench apparatus (Figure 5.18), three syringes are compressed by a common driving bar. The first two syringes contain enzyme and substrate solutions, respectively. These solutions flow into the first mixing chamber and then through a reaction aging tube to the second mixing chamber. The length of reaction time is controlled by the length of the reaction aging tube, or by the rate of flow through this tube. In the second mixing chamber, the reaction mixture is combined with a third solution containing the quenching material, which rapidly stops (or quenches) the reaction. The quenching solution must be able to instantaneously stop the reaction by denaturing the enzyme or sequestering a critical cofactor or other component of the reaction mixture. For example, strong acids are commonly used to quench enzyme reactions in this way. Also, enzymes that rely on divalent metal ions as necessary cofactors can be effectively quenched by mixing with EDTA (ethylenediaminetetraacetic acid) or other chelators. Once the reaction has been quenched, the mixed solution flows into a collection container, from which it can be retrieved by the scientist. Detection is performed off-line by any convenient method, including spectroscopy, radiometric chromatography (including thin-layer chromatography), or electrophoretic separation of substrates and products .

With instruments like stopped-flow and rapid quench apparatus, one can measure the formation of products or intermediates on a millisecond time scale. A number of kinetic schemes can be studied by these rapid kinetic techniques. We describe two common situations. The first is simple, reversible association of the subtrate and enzyme to form the ES complex:

$$E + S \underset{k_{-1}}{\overset{k_1}{\rightleftharpoons}} ES$$

If the experiment is performed under conditions where [S]>[E], formation of the ES complex will follow pseudo-first-order kinetics and will be equivalent to the approach to equilibrium for receptor—ligand complexes. Hence, the observed rate of formation k_{obs} will depend on substrate concentration as follows:

$$k_{obs} = k_1[S] + k_{-1} \tag{5.49}$$

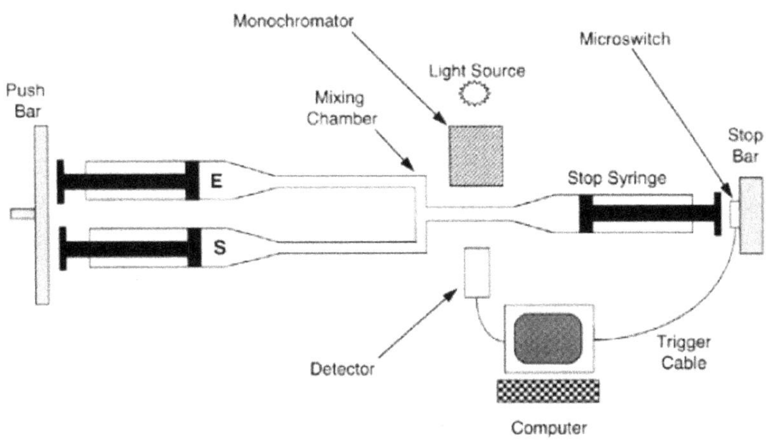

Figure 5.17 Schematic diagram of a typical stopped-flow instrument for rapid kinetic measurements.

A plot of k_{obs} as a function of [S] will be linear with y intercept of k_{-1}andslope of k_1 , as illustrated earlier (Figure 3.2).

In the second situation, formation of an intermediate species EX is rate limiting. Here initial substrate binding comes to equilibrium on a time scale much faster than the subsequent first-order "isomerization" step:

$$E + S \underset{k_{-1}}{\overset{k_1}{\rightleftharpoons}} ES \underset{k_{-2}}{\overset{k_2}{\rightleftharpoons}} EX$$

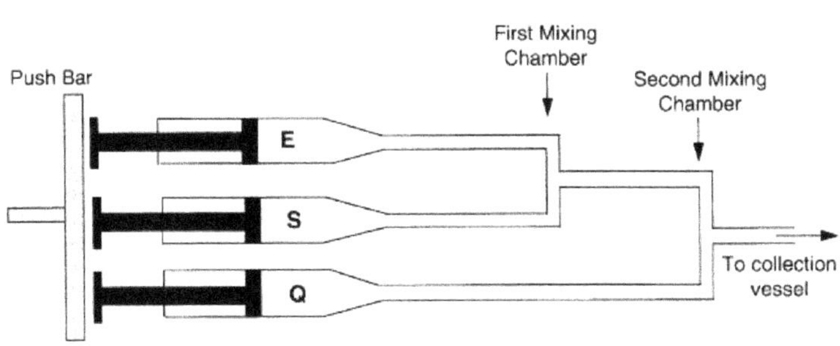

Figure 5.18 Schematic diagram of a typical rapid quench instrument for rapid kinetic measurements.

Three forms of the enzyme appear in this reaction scheme, the free enzyme E_f, the ES complex, and the intermediate EX. The time dependence for each of these can be derived, yielding the following equations for the mole fraction of each species :

$$\frac{[E]_f}{[E]} = \frac{1-(K_1[S]+K_1K_2[S])}{\alpha}(1-e^{\lambda t}) \tag{5.50}$$

$$\frac{[ES]}{[E]} = \frac{K_1[S]}{\alpha}(1-e^{\lambda t}) \qquad ((5.51)$$

$$\frac{[EX]}{[E]} = \frac{K_1K_2[S]}{\alpha}(1-e^{\lambda t}) \tag{5.52}$$

where

$$\alpha = 1 + K_1[S] + K_1K_2[S] \tag{5.53}$$

The preexponential term in Equations 5.50—5.52 represents in each case an amplitude term that corresponds to the concentration of that enzyme species at equilibrium. The observed rate constant for formation of the EX complex, λ., follows a hyperbolic dependence on substrate concentration, similar to velocity in the Michaelis—Menten equation:

$$\lambda = \frac{K_1k_2[S]}{K_1[S]+1} + k_{-2} \tag{5.54}$$

Three distinctions between Equations 5.54 and 5.24 can be made. First, λ is a hyperbolic function of the true dissociation constant for the ES complex (i.e., K_s=1/ K_1), not of the kinetic constant K_m. Second, the maximal rate observed is equal to the sum of k_2+k_{-2}. Third, the *y* intercept is nonzero in this case,and equal the rate constant k_{-2}. Thus, from fitting of the rapid kinetic data to Equation 5.54, one can simultaneously determine the values of K_s, k_2, and k_{-2}.

The application of transient kinetics to the study of enzymatic reactions, and more generally to protein—ligand binding events, is widespread throughout the biochemical literature. The reader should be aware of the power of these methods for determining individual rate constants and of the value of such information for the development of detailed mechanistic models of catalytic turnover. Because of space limitations, and because these methods require specialized equipment that beginners may not have at their disposal, we shall suspend further discussion of these methods. Several noteworthy reviews on the methods of transient kinetics are highly recommended to the reader who is interested in learning more about these techniques.

Chapter 6

CHEMICAL MECHANISMS IN ENZYME CATALYSIS

The essential role of enzymes in almost all physiological processes stems from two key features of enzymatic catalysis: (1) enzymes greatly *accelerate* the rates of chemical reactions; and (2) enzymes act on *specific* molecules, referred to as substrates, to produce *specific* reaction products. Together these properties of rate acceleration and substrate specificity afford enzymes the ability to perform the chemical conversions of metabolism with the efficiency and fidelity required for life. In this chapter we shall see that both substrate specificity and rate acceleration result from the precise three-dimensional structure of the substrate binding pocket within the enzyme molecule, known as the *active site*. Enzymes are (almost always) proteins, hence the chemically reactive groups that act upon the substrate are derived mainly from the natural amino acids. The identity and arrangement of these amino acids within the enzyme active site define the active site topology with respect to stereochemistry, hydrophobicity, and electrostatic character. Together these properties define what molecules may bind in the active site and undergo catalysis. The active site structure has evolved to bind the substrate molecule in such a way as to induce strains and perturbations that convert the substrate to its transition state structure. This transition state is greatly stabilized when bound to the enzyme; its stability under normal solution conditions is much less. Since attainment of the transition state structure is the main energetic barrier to the progress of any chemical reaction, we shall see that the stabilization of the transition state by enzymes results in significant acceleration of the reaction rate.

6.1 SUBSTRATE--ACTIVE SITE COMPLEMENTARITY

When a protein and a ligand combine to form a binary complex, the complex must result in a net stabilization of the system relative to the free protein and ligand; otherwise binding would not be thermodynamically favorable. We discussed in another Chapter the main forces involved in stabilizing protein—ligand interactions: hydrogen bonding, hydrophobic forces, van der Waals interactions, electrostatic interactions, and so on. All these contribute to the overall binding energy of the complex and must more than compensate for the lose of rotational and translational entropy that accompanies binary complex formation.

These same forces are utilized by enzymes in binding their substrate molecules. It is clear today that formation of an enzyme—substrate binary complex is but the first step in the catalytic process used in enzymatic catalysis. Formation of the initial encounter complex (also referred to as the enzyme— substrate, ES, or Michaelis complex;) is followed by steps leading sequentially to a stabilized enzyme—transition state complex (ES‡), an enzyme— product complex (EP), and finally dissociation to reform the free enzyme with liberation of product molecules. Initial ES complex formation is defined by the dissociation constant K_s , which is the quotient of the rate constants k_{on} and k_{off} . As discussed in Chapter 5, the rates of the chemical steps following ES complex formation are, for simplicity, often collectively described by a single kinetic constant, k_{ca}t. As we shall see, k_{cat} is most often limited by the rate of attainment of the transition state species ES‡. Hence, a minimalist view of enzyme catalysis is captured in the scheme illustrated in Figure 6.1.

To understand the rate enhancement and specificity of enzymatic reactions, we must consider the structure of the reactive center of these molecules, the active site, and its relationship to the structures of the substrate molecule in its ground and transition states in forming the ES and the ES‡ binary complexes.

While the active site of every enzyme is unique, some generalizations can be made:

1. The active site of an enzyme is small relative to the total volume of the enzyme.
2. The active site is three-dimensional—that is, amino acids and cofactors in the active site are held in a precise arrangement with respect to one another and with respect to the structure of the substrate molecule. This active site three-dimensional structure is formed as a result of the overall tertiary structure of the protein.
3. In most cases, the initial interactions between the enzyme and the substrate molecule (i.e., the binding events) are noncovalent, making use of hydrogen bonding, electrostatic, hydrophobic interactions, and van der Waals forces to effect binding.

$$E + S \underset{K_S}{\overset{\Delta G_{ES}}{\rightleftharpoons}} ES \xrightarrow[\Delta G_{kcat}]{k_{cat}} E + P$$

$$E + S \overset{k_{cat}/K_S}{\rightleftharpoons} ES^{\ddagger}$$

$$E_a = \Delta G_{ES\ddagger}$$

Figure 6.1 Generic scheme for an enzyme-catalyzed reaction showing the component free energy terms that contribute to the overall activation energy of reaction.

4. The active sites of enzymes usually occur in clefts and crevices in the protein. This design has the effect of excluding bulk solvent (water), which would otherwise reduce the catalytic activity of the enzyme. In other words, the substrate molecule is desolvated upon binding, and shielded from bulk solvent in the enzyme active site. Solvation by water is replaced by the protein.
5. The specificity of substrate utilization depends on the well-defined arrangement of atoms in the enzyme active site that in some way complements the structure of the substrate molecule.

Experimental evidence for the existence of a binary ES complex rapidly accumulated during the late nineteenth and early twentieth centuries. This evidence, some of which was discussed in Chapter 5, was based generally on studies of enzyme stability, enzyme inhibition, and steady state kinetics. During this same time period, scientists began to appreciate the selective utilization of specific substrates that is characteristic of enzyme-catalyzed reactions. This cumulative information led to the general view that substrate specificity was a result of selective binding of substrate molecules by the enzyme at its active site.

The selection of particular substrates reflected a structural complementarity between the substrate molecule and the enzyme active site. In the late nineteenth century Emil Fisher formulated these concepts into the *lock and key* model, as illustrated in Figure 6.2. In this model the enzyme active site and the substrate molecule are viewed as static structures that are stereochemically complementary. The insertion of the substrate into the static enzyme active site is analogous to a key fitting into a lock, or a jigsaw piece fitting into the rest of the puzzle: the best fits occur with the substrates that best complement the structure of the active site; hence these molecules bind most tightly.

Active site—substrate complementarity results from more than just stereochemical fitting of the substrate into the active site. The two structures must also be electrostatically complementary, ensuring that charges are counterbalanced to avoid repulsive effects. Likewise, the two structures must complement each other in the arrangement of hydrophobic and hydrogenbonding interactions to best enhance binding interactions.

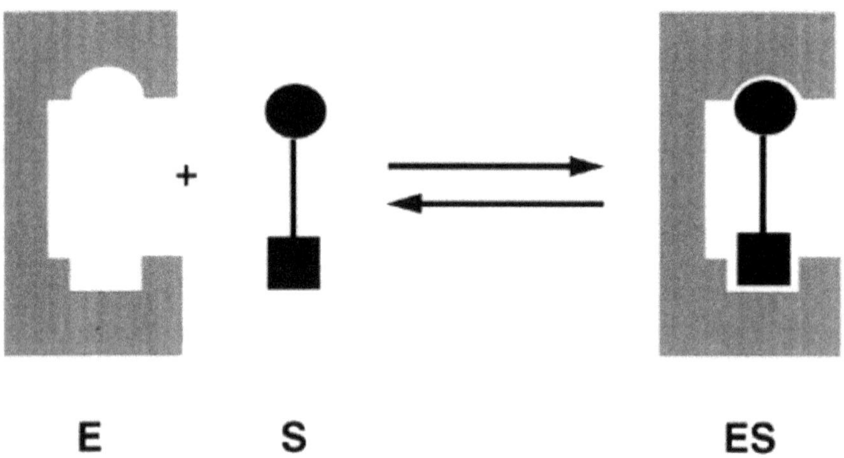

Figure 6.2 Schematic illustration of the lock and key model of enzyme—substrate interactions.

Enzyme catalysis is usually stereo-, regio-, and enantiomerically selective. Hence substrate recognition must result from a minimum of three contact points of attachment between the enzyme and the substrate molecule. Consider the example of the alcohol dehydrogenases that catalyze the transfer of a methylene hydrogen of ethyl alcohol to the carbon at the 4-position of the NAD^+ cofactor, forming NADH and acetaldehyde. Studies in which the methylene hydrogens of ethanol were replaced by deuterium demonstrated that alcohol dehydrogenases exclusively transferred the pro-*R* hydrogen to NAD^+ .This stereospecificity implies that the alcohol bind to the enzyme active site through specific interactions of its methyl, hydroxyl, and pro-*R* hydrogen groups to form a *three-point attachment* with the reactive groups within the active site; this concept is illustrated in Figure 6.3. Having anchored down the methyl and hydroxyl groups as depicted in Figure 6.3, the enzyme is committed to the transfer of the specific pro-*R* hydrogen atom because of its relative proximity to the NAD^+ cofactor. The three-point attachment hypothesis is often invoked to explain the stereospecificity commonly displayed by enzymatic reactions.

The concepts of the lock and key and three-point attachment models help to explain substrate selectivity in enzyme catalysis by invoking a structural complementarity between the enzyme active site and substrate molecule. We have not, however, indicated the form of the substrate molecule to which the enzyme active site shows structural complementarity. Early formulations of these hypotheses occurred before the development of transition state theory , hence viewed the substrate ground state as the relevant configuration. Today, however, there is clear evidence that enzyme active sites have in fact evolved to best complement the substrate transition state structure, rather than the ground state. For example, it is well known that inhibitor molecules that are designed to mimic the structure of the reaction transition state bind much more tightly to the target enzyme than do the substrate or product molecules. Some scientists have, in fact, argued that "the sole source of catalytic power is the stabilization of the transition state; reactant state interactions are by nature inhibitory and only waste catalytic power" . Others argue that some substrate ground state affinity is required for initial complex formation and to utilize the accompanying binding energy to drive transition state formation. Indeed some evidence from site-directed mutagenesis studies suggests that the structural determinants of substrate specificity can at least in part be distinguished from the mechanism of transition state stabilization . Nevertheless, the bulk of the experimental evidence strongly favors active site—transition state complementarity as the primarybasis for substrate specificity and catalytic power in most enzyme systems. There are, for example, numerous studies of specificity in enzyme systems measured through steady state kinetics in which specificity is quantified in terms of the relative k_{cat}/K_m values for different substrates. In many of these studies one finds that the K_m values among different substrates vary very little, perhaps by a factor of 10-fold or less. A good substrate is distinguished from a bad one in these studies mainly by the effects on K_{cat}. Hence, much of the substrate specificity resides in transition state interactions with the enzyme active site.

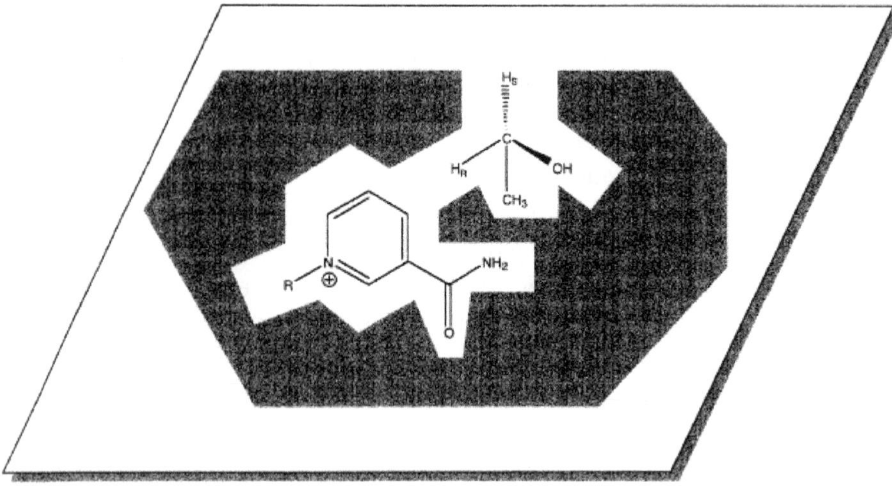

Figure 6.3 Illustration of three-point attachment in enzyme—substrate interactions.

6.2 RATE ENHANCEMENT THROUGH TRANSITION STATE STABILIZATION

we said that chemical reactions, such as molecule S (for substrate) going to molecule P (for product), will proceed through formation of a high energy, short-lived (typical half-life ca. 10^{-13} second) state known as the transition state (S‡). Let us review the minimal steps involved in catalysis, as illustrated in Figure 6.1. The initial encounter (typically through molecular collisions in solution) between enzyme and substrate leads to the reversible formation of the Michaelis complex, ES. Under typical laboratory conditions this equilibrium favors formation of the complex, with ΔG of binding for a typical ES pair being approximately -3 to -12 kcal/mol. Formation of the ES complex leads to formation of the bound transition state species ES‡. As with the uncatalyzed reaction, formation of the transition state species is the main energetic barrier to product formation. Once the transition state barrier has been overcome, the reaction is much more likely to proceed energetically downhill to formation of the product state. In the case of the enzyme-catalyzed reaction, this process involves formation of the bound EP complex, and finally dissociation of the EP complex to liberate free product and free enzyme. Since the enzyme appears on both the reactant and product side of the equation and is therefore unchanged with respect to the thermodynamics of the complete reaction, it can be ignored . Hence, the free energy of the reaction here will depend only on the relative concentrations of S and P:

$$\Delta G = -RT\ln\left(\frac{[P]}{[S]}\right) \tag{6.1}$$

This is exactly the same equation of ΔG for the uncatalyzed reaction of S->P, and it reflects the *path independence* of the function ΔG. In other words, ΔG depends only on the initial and final states of the reaction, not on the various intermediate states (e.g., ES, ES‡, and EP) formed during the reaction (ΔG is thus said to be a *state function*). This leads to the important realization that *enzymes cannot alter the equilibrium between products and substrates.*

What then is the value of using an enzyme to catalyze a chemical reaction?

The answer is that enzymes, and in fact all catalysts, speed up the rate at which equilibrium is established in a chemical system: *enzymes accelerate the rate of chemical reactions.* Hence, with an ample supply of substrate, one can form much greater amounts of product *per unit time* in the presence of an enzyme than in its absence. This rate acceleration is a critical feature of enzyme usage in metabolic processes. Without the speed imparted by enzyme catalysis, many metabolic reactions would proceed too slowly in vivo to sustain life. Likewise, the *ex vivo* use of enzymes in chemical processes relies on this rate acceleration, as well as the substrate specificity that enzyme catalysis provides. Thus, the great value of enzymes, both for biological systems and in commercial use, is that they provide a means of making more product at a faster rate than can be achieved without catalysis.

How is it that enzymes achieve this rate acceleration? The answer lies in a consideration of the activation energy of the chemical reaction. The key to enzymatic rate acceleration is that by lowering the energy barrier, by stabilizing the transition state, reactions will proceed faster.

Recall from earlier Chapter that the rate or velocity of substrate utilization, *v*, is related to the activation energy of the reaction as follows:

$$v = \frac{-d[S]}{dt} = \left(\frac{k_B T}{h}\right)[S]\exp\left(-\frac{E_a}{RT}\right) \tag{6.2}$$

Now, for simplicity, let us fix the reaction temperature at 25°C and fix [S] at a value of 1 in some arbitrary units. At 25°C, RT =0.59 and K_BT/h=6.2_X10^{12} s^{-1}. Suppose that the activation energy of a chemical reactionat 25°C is 10 kcal/mol. The velocity of the reaction will thus be:

$$v = 6.2 \times 10^{12}\,s^{-1} \cdot 1\,\text{unit} \cdot e^{-10/0.59} = 2.7 \times 10^{5}\ \text{units}\,s^{-1} \tag{6.3}$$

If we somehow reduce the activation energy to 5 kcal/mol, the velocity now becomes:

$$v = 6.2 \times 10^{12}\,\mathrm{s}^{-1} \cdot 1\,\mathrm{unit} \cdot e^{-5/0.59} = 1.3 \times 10^{9}\,\mathrm{units\,s}^{-1} \tag{6.4}$$

Thus by lowering the activation energy by 5 kcal/mol we have achieved an increase in reaction velocity of about 5000! In general a linear decrease in activation energy results in an exponential increase in reaction rate. This is exactly how enzymes function. They accelerate the velocity of chemical reactions by stabilizing the transition state of the reaction, hence lowering the energetic barrier that must be overcome.

Let us look at the energetics of a chemical reaction in the presence and absence of an enzyme. For the enzyme-catalyzed reaction we can estimate the free energies associated with different states from a combination of equilibrium and kinetic measurements. If we normalize the free energy of the free ESstarting point to zero, we can calculate the free energy change associated with ES‡ (under experimental conditions of subsaturating substrate) as follows:

$$E_a = \Delta G_{ES^\ddagger} = -RT\ln\left(\frac{k_{cat}}{K_s}\right) + RT\ln\left(\frac{k_B T}{h}\right) \tag{6.5}$$

The free energy change associated with formation of the ES complex can, in favorable cases, be determined from measurement of K_s by equilibrium methods or from kinetic measurements :

$$\Delta G_{ES} = -RT\ln\left(\frac{1}{K_s}\right) \tag{6.6}$$

Alternatively, from steady state measurements one can calculate the free energy change associated with k_{cat} from the Eyring equation:

$$\Delta G_{k_{cat}} = RT\left(\ln\left(\frac{k_B T}{h}\right) - \ln\left(k_{cat}\right)\right) \tag{6.7}$$

If one then subtracts Equation 6.7 from Equation 6.5, the difference is equal to ΔG_{ES}. Thus, we see that the overall activation energy E_a is composed of two terms, ΔG_{ES} and $\Delta G_{k_{cat}}$. The term $\Delta G_{k_{cat}}$ is the amount of energy that must be expended to reach the transition state (i.e., bond-making and bond-breakingis the net energy gain that results from the realization of enzyme—substrate binding energy .

The free energy change associated with the EP complex can also be determined from equilibrium measurements or from the inhibitory effects of product on the steady state kinetics of the reaction .

For either the catalyzed or uncatalyzed reaction, the activation energy can also be determined from the temperature dependence of the reaction velocity according to the Arrhenius equation :

$$k_{cat} = A\exp\left(\frac{-E_a}{RT}\right) \tag{6.8}$$

Note that for the uncatalyzed reaction, k_{cat} is replaced in Equation 6.8 by the first-order rate constant for reaction.

From such measurements one can construct a reaction energy level diagram as illustrated in Figure 6.4. In the absence of enzyme, the reaction proceeds from substrate to product by overcoming the sizable energy barrier required to reach the transition state S‡. In the presence of enzyme, on the other hand, the reaction first proceeds through formation of the ES complex. The ES complex represents an intermediate along the reaction pathway that is not available in the uncatalyzed reaction; the binding energy associated with ES complex formation can, in part, be used to drive transition state formation.

Once binding has occurred, molecular forces in the bound molecule (as discussed shortly) have the effect of simultaneously destabilizing the ground state configuration of the bound substrate molecule, and energetically favoring the transition state. The complex ES‡ thus occurs at a lower energy than the free S‡ state, as shown in Figure 6.4.

The reaction next proceeds through formation of another intermediate state, the enzyme—product complex, EP, before final product release to form the free product plus free enzyme state. Again, the initial and final states are energetically identical in the catalyzed and uncatalyzed reactions. However, the overall activation energy barrier has been substantially reduced in the enzymecatalyzed case. This reduction in activation barrier results in a significant acceleration of reaction velocity in the presence of the enzyme, as we have seen above (Equations 6.2—6.4). This is the common strategy for rate acceleration used by all enzymes:

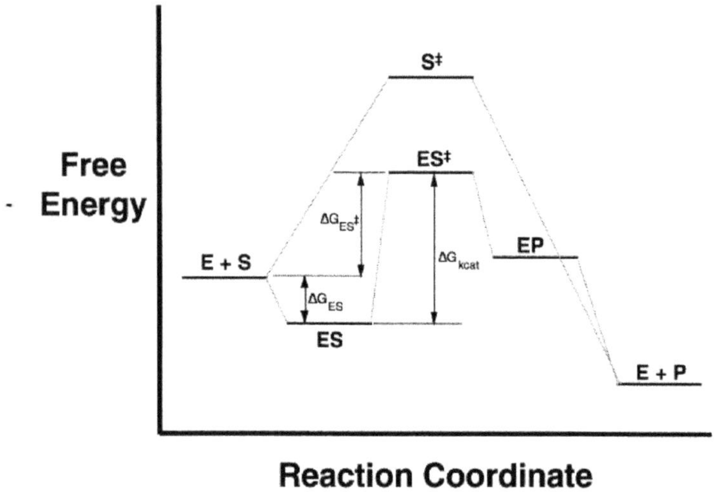

Figure 6.4 Energy level diagram of an enzyme-catalyzed reaction and the corresponding uncatalyzed chemical reaction. The symbols E, S, S‡, ES, ES‡, EP, and P represent the free enzyme, the free substrate, the free transition state, the enzyme–substrate complex, the enzyme–transition state complex, the enzyme–product complex, and the free product states, respectively. The activation energy, $\Delta G_{ES‡}$ and its components, ΔG_{ES} and $\Delta G_{k_{cat}}$, are as described in the text. The energy levels depicted relate to the situation in which the substrate is present in concentrations greater than the dissociation constant for the ES complex. When [S] is less than K_S, the potential energy of the ES state is actually greater than that of the E+S initial state .

6.3 CHEMICAL MECHANISMS FOR TRANSITION STATE STABILIZATION

The transition state stabilization associated with enzyme catalysis is the result of the structure and reactivity of the enzyme active site, and how these structural features interact with the bound substrate molecule. Enzymes use numerous detailed chemical mechanisms to achieve transition state stabilization and the resulting reaction rate acceleration. These can be grouped into five major categories :

1. Approximation (i.e., proximity) of reactants
2. Covalent catalysis
3. General acid—base catalysis
4. Conformational distortion
5. Preorganization of the active site for transition state complementarity

We shall discuss each of these separately. However, the reader should realize that in any catalytic system, several or all of these effects can be utilized in concert to achieve overall rate enhancement. They are thus often interdependent, which means that the line of demarcation between one mechanism and another often is unclear, and to a certain extent arbitrary.

6.3.1 Approximation of Reactants

Several factors associated with simply binding the substrate molecule within the enzyme active site contribute to rate acceleration. One of the more obvious of these is that binding brings into close proximity (hence the term *approximation*), the substrate molecule(s) and the reactive groups within the enzyme active site. Let us consider the example of a bimolecular reaction, involving two substrates, A and B, that react to form a covalent species A—B. For the two molecules to react in solution they must (1) encounter each other through diffusion-limited collisions in the correct mutual orientations for reaction; (2) undergo changes in solvation that allow for molecular orbital interactions; (3) overcome van der Waals repulsive forces; and (4) undergo changes in electronic orbitals to attain the transition state configuration.

In solution, the rate of reaction is determined by the rate of encounters between the two substrates. The rate of collisional encounters can be marginally increased in solution by elevating the temperature, or by increasing the concentrations of the two reactants. In the enzyme-catalyzed reaction, the two substrates bind to the enzyme active site as a prerequisite to reaction. When the substrates are sequestered within the active site of the enzyme, their *effective* concentrations are greatly increased with respect to their concentrations in solution.

A second aspect of approximation effects is that the structure of the enzyme active site is designed to bind the substrates in a specific orientation that is optimal for reaction. In most bimolecular reactions, the two substrates must achieve a specific mutual alignment to proceed to the transition state. In solution, there is a distribution of rotomer populations for each substrate that have the effect of retarding the reaction rate. By locking the two substrates into a specific mutual orientation in the active site, the enzyme overcomes these encumberances to transition state attainment. Of course, these severe steric and orientational restrictions are associated with some entropic cost to reaction.

However, such alignment must occur for reaction in solution as well as in the enzymatic reaction. Hence, there is actually a considerable entropic advantage associated with reactant approximation. By having the two substrates bound in the enzyme active site, the entropic cost associated with the solution reaction is largely eliminated; in enzymatic catalysis this energetic cost is compensated for in terms of the binding energy of the ES complex. Together, the concentration and orientation effects associated with substrate binding are referred to as the *proximity effect* or the *propinquity effect.*

Some sense of the effects of proximity on reaction rate can be gleaned from studies comparing the reaction rates of intramolecular reactions with those of comparable intermolecular reactions . For example, Fersht and Kirby compared the reaction rate of aspirin hydrolysis catalyzed by the intramolecular carboxylate group with that for the same reaction catalyzed by acetate ions in solution .

The intramolecular reaction proceeds with a first-order rate constant of $1.1X10^{-5}$ s^{-1}. The same reaction catalyzed by acetate ions in solution has a second-order rate constant of $1.27X10^{-6}Lmol^{-1}\ s^{-1}$. In comparing these two reactions we can ask what effective molarity of acetate ions would be required to make the intermolecular reaction go at the same rate as the intramolecular reaction. This is measured as the ratio of the first-order rate constant to the second-order rate constant (k_1/k_2); this ratio has units of molarity, and its value for the present reaction is 8.7 M. However, becauseacetate is more basic (pK_a 4.76) than the carboxylate of aspirin (pK_a 3.69), one must adjust the value of *k*-to account for this difference. When this is done,the effective molarity is 13 M. Thus, with the pK_a adjustment corrected for, theoverall rate of the intramolecular reaction is far greater than that of theintermolecular reaction.

A concept related to proximity effects is that of *orbital steering.* The orbital steering hypothesis suggests that the juxtaposition of reactive groups among the substrates and active site residues is not sufficient for catalysis. In addition to this positioning, the enzyme needs to precisely steer the molecular orbitals of the substrate into a suitable orientation. According to this hypothesis, enzyme active site groups have evolved to optimize this steering upon substrate binding. While some degree of orbital steering no doubt occurs in enzyme catalysis, there are two strong arguments against a major role for this effect in transition state stabilization:

1. Thermal vibrations of the substrate molecules should give rise to large changes in the orientation of the reacting atoms within the active site structure. The magnitude of such vibrational motions at physiological temperatures contradicts the idea of rigidly oriented molecular orbitals as required for orbital steering.
2. Recent molecular orbital calculations predict that orbital alignments result in shallow total energy minima , whereas the orbital steering hypothesis would require deep, narrow energy minimal to retain the exact alignment. An expanded discussion of orbital steering and the arguments for and against this hypothesis has been provided by Bender et al. .

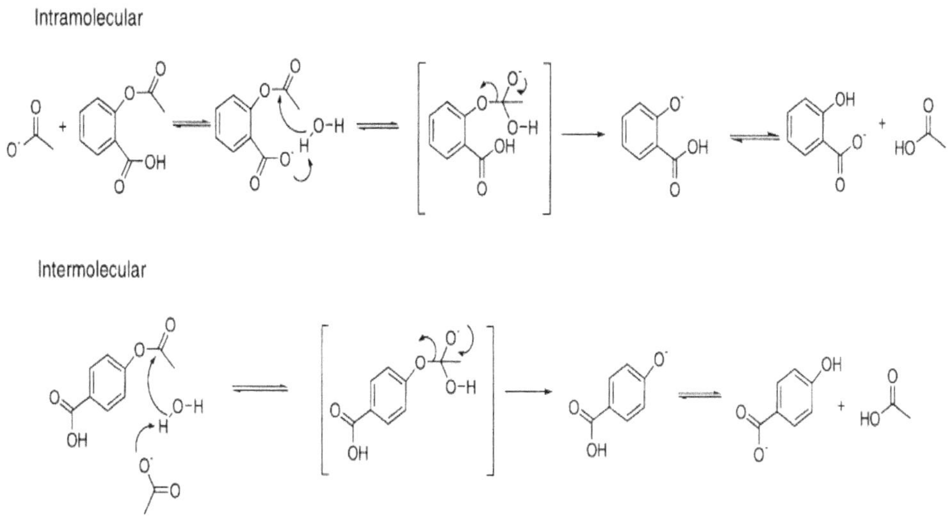

Scheme 6.1

Changes in solvation are also required for reaction between two substrates to occur. In solution, desolvation energy can be a large barrier to reaction. In enzymatic reactions the desolvation of reactants occurs during the binding of substrates to the hydrophobic enzyme active site, where they are effectively shielded from bulk solvent. Hence desolvation costs are offset by the binding energy of the complex and do not contribute to the activation barrier in the enzymatic reaction .

Finally, overcoming van der Waals repulsions and changes in electronic overlap are important aspects of intramolecular reactions and enzyme catalysis. These ends are accomplished in part by the orientation effects discussed above, and through induction of strain, as discussed latter in this chapter. Together these different properties lead to an overall approximation effect that results from the binding of substrates in the enzyme active site.

Approximation effects contribute to the overall rate acceleration seen in enzyme catalysis, with the binding forces between the enzyme and substrate providing much of the driving force for these effects.

6.3.2 Covalent Catalysis

There are numerous examples of enzyme-catalyzed reactions that go through the formation of a covalent intermediate between the enzyme and the substrate molecule. Experimental evidence for such intermediates has been obtained from kinetic measurements, from isolation and identification of stable covalent adducts, and more recently from x-ray crystal structures of the intermediate species. Several families of enzymes have been demonstrated to form covalent intermediates, including serine proteases (acyl—serine intermediates), cysteine proteases (acyl—cysteine intermediates), protein kinases and phosphatase (phospho—amino acid intermediates), and pyridoxal phosphate-utilizing enzymes (pyridoxal—amino acid Schiff bases).

For enzymes that proceed through such mechanisms, formation of the covalent adduct is a required step for catalysis. Generation of the covalent intermediate brings the system along the reaction coordinate toward the transition state, thus helping to overcome the activation energy barrier.

Enzymes that utilize covalent intermediates have evolved to break this difficult reaction down into two steps—formation and breakdown of the covalent intermediate—rather than catalysis of the single reaction directly. The ratelimiting step in the reactions of these enzymes is often the formation or decomposition of the covalent intermediate. This can be seen, for example, in Figure 6.5, which illustrates the steady state kinetics of *p*-nitrophenylethyl carbonate hydrolysis by chymotrypsin. Figure 6.5 shows the steady state progress curves for hydrolysis at several different enzyme concentrations. For an experiment of this type, one would expect the steady state rate to increase with enzyme concentration (i.e., the slopes of the lines should increase with increasing enzyme), but all the curves should converge at zero product concentration at zero time. Instead, one sees in Figure 6.5 that the *y* intercept for each progress curve is nonzero, and the value of the intercept increases with enzyme concentration. In fact, extrapolation of the steady state lines to time zero results in a *y* intercept equal to the concentration of enzyme active sites present in solution. The early "burst" in product formation is the result of a single turnover of the enzyme with substrate, and formation of a stoichiometric amount of acyl—enzyme intermediate and product. Formation of the acyl intermediate in this case is fast, but the subsequent decomposition of the intermediate is rate limiting. Since further product formation cannot proceed without decomposition of the acyl intermediate, a burst of rapid kinetics is observed, followed by a much slower steady state rate of catalysis.

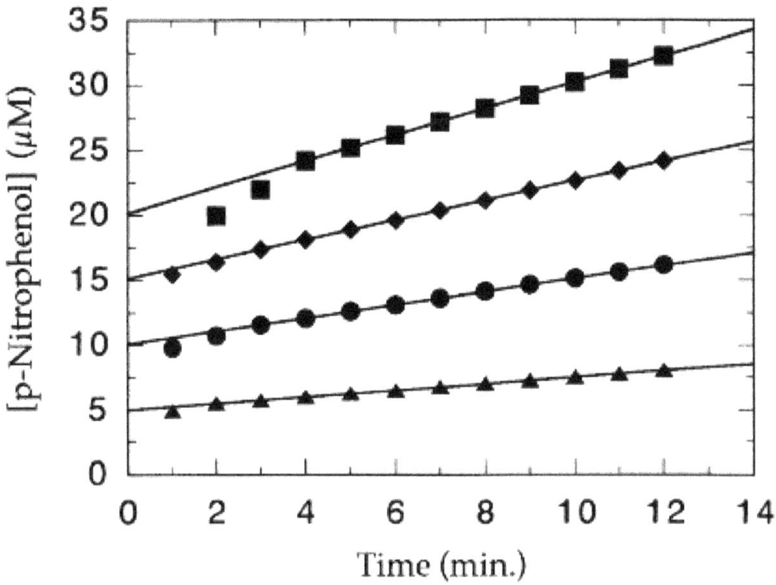

Figure 6.5 Illustration of burst phase kinetics. The data represent production of p-nitrophenol from the chymotrypsin-catalyzed hydrolysis of p-nitrophenylethyl carbonate. The nominal chymotrypsin concentrations used were 8 (triangles), 16 (circles), 24 (diamonds), and 32 (squares) μM. From the intercept values, the fraction of active enzyme in these samples was estimated to be 0.63. Note the apparent curvature in the early time points at high enzyme concentration, demonstrating a pre–steady state phase (i.e., the burst) in these reactions.

6.3.2.1 Nucleophilic Catalysis

Nucleophilic reactions involve donation ofelectrons from the enzyme nucleophile to a substrate with partial formation of a covalent bond between the groups in the transition state of the reaction:

$$\text{Nuc:} + \text{Sub} \leftrightharpoons \overset{\delta+}{\text{Nuc}} \text{---} \overset{\delta+}{\text{Sub}}$$

The reaction rate in nucleophilic catalysis depends both on the strength o the attacking nucleophile and on the susceptibility of the substrate group (electrophile) that is being attacked (i.e., how good a "leaving group" the attacked species has). The electron-donating ability, or *nucleophilicity*, of a group is determined by a number of factors; one of the most important of these factors is the basicity of the group. Basicity is a measure of the tendency of a species to donate an electron pair to a proton, as discussed in Chapter 2 and further in Section 6.3.3. Generally, the rate constant for reaction in nucleophilic catalysis is well correlated with the pK_a of the nucleophile. A plot of the logarithm of the second-order rate constant (k_2) of nucleophilic reaction as a function of nucleophile pK_a yields a straight line within a family of structurally related nucleophiles (Figure 6.6). A graph such as Figure 6.6 is known as a Brønsted plot because it was first used to relate the reaction rate to catalyst pK_a in general acid—base catalysis.

Note that in Figure 6.6 different structural families of nucleophiles all yield linear Brønsted plots, but with differing slopes, depending on the chemical nature of the nucleophile. Other factors that affect the strength of a nucleophile include oxidation potential, polarizability, ionization potential, electronegativity, potential energy of its highest occupied molecular orbital (HOMO); covalent bond strength, and general size of the group. Hence the reaction rate for nucleophilic catalysis depends not just on the pK_a of the nucleophile but also on the chemical nature of the species . This is one property that distinguishes nucleophilic catalysis from general base catalysis. While the Brønsted plot slope depends on the nature of the nucleophilic species in nucleophilic catalysis, in general base catalysis the slope depends solely on catalyst pK_a.

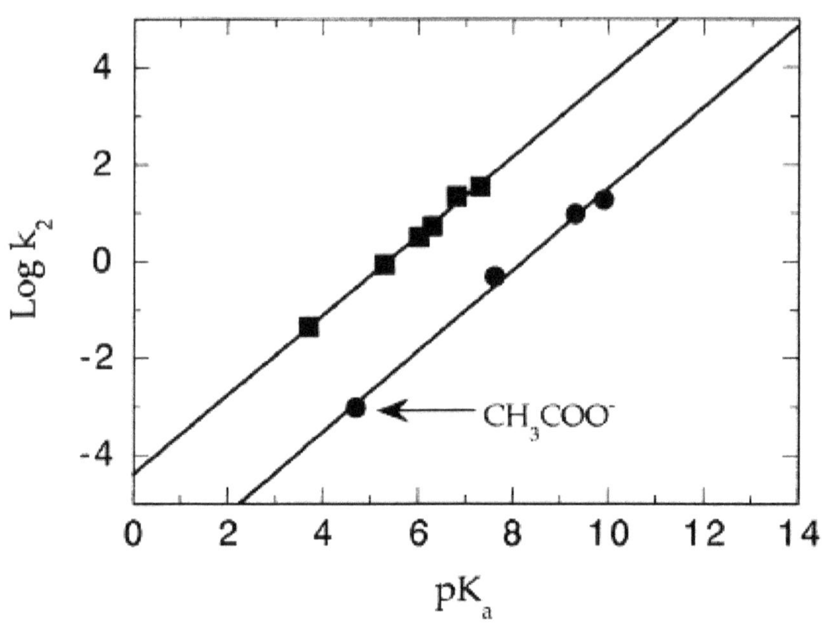

Figure 6.6 Brønsted plots for nucleophilic attack of p-nitrophenyl acetate by imidazoles (squares) and phenolates (circles). Unlike general base catalysis,

in this illustration N- and O-containing nucleophiles of similar basicity (pK_a) show distinct Brønsted lines. Note that the data points for acetate ion and the phenolate nucleophiles fall on the same Brønsted line

The most distinguishing feature of nucleophilic catalysis, however, is the formation of a stable covalent bond between the nucleophile and substrate along the path to the transition state. Often these covalent intermediates resemble isolable reactive species that are common in small molecule organic chemistry. This and other distinctions between nucleophilic and general base catalysis are presented in Section 6.3.3.

The susceptibility of the electrophile is likewise affected by several factors. Again, the pK_a of the leaving group, hence its state of protonation, appears to be a dominant factor. Studies of the rates of catalysis by a common nucleophile on a series of leaving groups demonstrate a clear correlation between rate of attack and the pK_a of the leaving group; generally, the weaker the base, the better leaving group the species. As with the nucleophile itself, the chemical nature of the leaving group, not its pK_a alone, also affects catalytic rate. Other factors influencing the ability of a group to leave can be found in the texts by Fersht and Jencks , and in most general physical organic chemistry texts.

In enzymatic nucleophilic catalysis, the nucleophile most often is an amino acid side chain within the enzyme active site. From the preceding discussion, one might expect the most basic amino acids to be the best nucleophiles in enzymes. Enzymes, however, must function within a narrow physiological pH range (around pH 7.4), and this limits the correlation between pK_a and nucleophilicity just described. For example, referring to Table 6.1, we might infer that the guanidine group of arginine would be a good nucleophile (pK_a 12.5). Consideration of the Henderson—Hasselbalch equation , however, reveals that at physiological pH (ca. 7.4) this group would exist almost entirely in the protonated conjugate acid form. Hence, arginine side chains do not generally function as nucleophiles in enzyme catalysis.

The amino acids that are capable of acting as nucleophiles are serine, threonine, cysteine, aspartate, glutamate, lysine, histidine, and tyrosine. Examples of some enzymatic nucleophiles and the covalent intermediates they form are given in Table 6.1. A more comprehensive description of nucleophilic catalysis and examples of its role in enzyme mechanisms can be found in the text by Walsh .

6.3.2.2 Electrophilic Catalysis

In electrophilic catalysis covalent intermediates are also formed between the cationic electrophile of the enzyme and an electron-rich portion of the substrate molecule. The amino acid side chains do not provide very effective electrophiles.

Hence, enzyme electrophilic catalysis most often require electron-deficient organic cofactors or metal ions.

Table 6.1 Some examples of enzyme nucleophiles and the covalent intermediates formed in their reactions with substrates

Nucleophilic Group	Example Enzyme	Covalent Intermediate
Serine (—OH)	Serine proteases	Acyl enzyme
Cysteine (—SH)	Thiol proteases	Acyl enzyme
Aspartate (—COO^-)	ATPases	Phosphoryl enzyme
Lysine (—NH_2)	Pyridoxal-containing enzymes	Schiff bases
Histidine	Phosphoglycerate mutase	Phosphoryl enzyme
Tyrosine (—OH)	Glutamine synthase	Adenyl enzyme

There are numerous examples of enzymatic reactions involving active sitemetal ions in electrophilic catalysis. Themetal can play a number of possible roles in these reactions: it can shield negative charges on substrate groups that would otherwiserepel attack of an electron pair from a nucleophile; it can act to increase the reactivity of a group by electron withdrawal; and it can act to bridge a substrate and nucleophilic group; they can alter the pK_a and reactivity of nearby nucleophiles.

Metal ions are also used in enzyme catalysis as binding centers for substrate molecules. For example, in many of the cytochromes, the heme iron ligates the substrate at one of its six coordination sites and facilitates electron transfer to the substrate. Metal ions bound to substrates can also affect the substrate conformation to enhance catalysis; that is, they can change the geometry of a substrate molecule in such a way as to facilitate reactivity. In the ATPdependent protein kinases (enzymes that transfer a γ-phosphate group from ATP to a protein substrate), for example, the substrate of the enzyme is not ATP itself, but rather an ATP—Mg^{2+} coordination complex. The Mg^{2+} binds at the terminal phosphates, positioning these groups to greatly facilitate nucleophilic attack on the γ-phosphate.

The most common mechanism of electrophilic catalysis in enzyme reactions is one in which the substrate and the catalytic group combine to generate, in situ, an electrophile containing a cationic nitrogen atom. Nitrogen itself is not a particularly strong electrophile, but it can act as an effective electron sink in such reactions because of its ease of protonation and because it can form cationic unsaturated adducts with ease. A good example of this is the family of electrophilic reactions involving the pyridoxal phosphate cofactor.

Pyridoxal phosphate is a required cofactor for the majority of enzymes catalyzing chemical reactions at the alpha, beta, and gamma carbons of the α-amino acid.

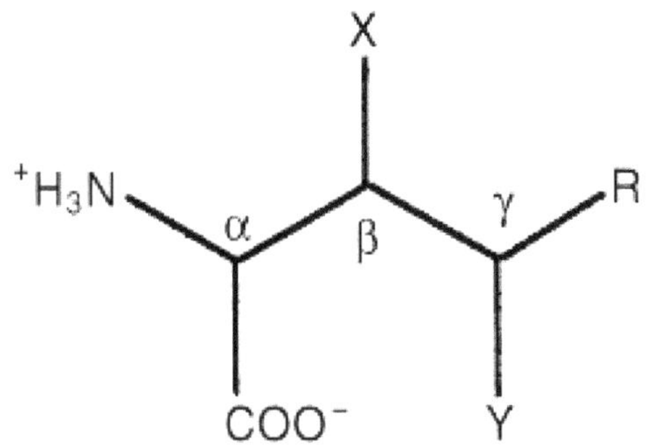

The first step in reactions between the amino acids and the cofactor is the formation of a cationic imine (Schiff base), which plays a key role in lowering the activation energy.

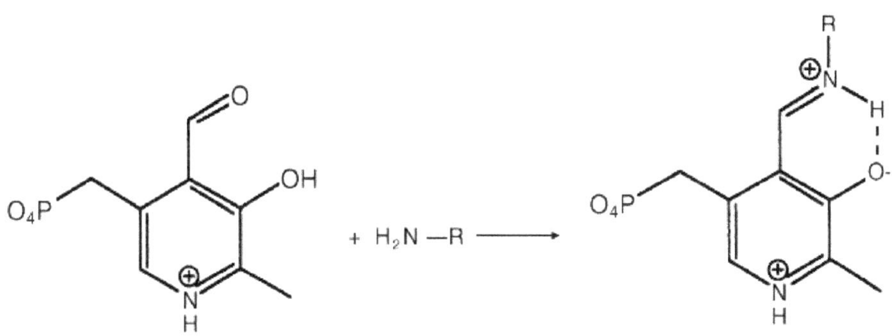

Pyridoxal phosphate Amine Schiff base

The function of the pyridoxal phosphate here is to act as an electron sink, stabilizing the carbanion intermediate that forms during catalysis. Electron withdrawal from the alpha-carbon of the attached amino acid toward the cationic nitrogen activates all three substituents for reaction; hence any one of these can be cleaved to form an anionic center. Because the cationic imine is conjugated to the heteroaromatic pyridine ring, significant charge delocalization is provided, thus making the pyridoxal phosphate group a very efficient catalyst for electrophilic reactions.

All pyridoxal-containing enzymes proceed through three basic steps: formation of the cationic imine, chemical changes through the carbanion intermediate, and hydrolysis of the product imine. A common reaction of pyridoxal phosphate with α-amino acids is removal of the α-hydrogen (Figure 6.7) to give a key intermediate in a variety of amino acid reactions, including transamination, racemization, decarboxylation, and side chain interconversion . In transamination (Figure 6.7), for example, removal of the α-hydrogen is followed by proton donation to the pyridoxal phosphate carbonyl carbon, leading to formation of an α-keto acid—pyridoxamine Schiff base. Subsequent hydrolysis of this species yields the free α-keto acid and the pyridoxamine group. An imine is then formed between the keto acid and pyridoxamine, and reversed proton transfer occurs to generate a new amino acid and to regenerate the pyridoxal, thus completing the catalytic cycle.

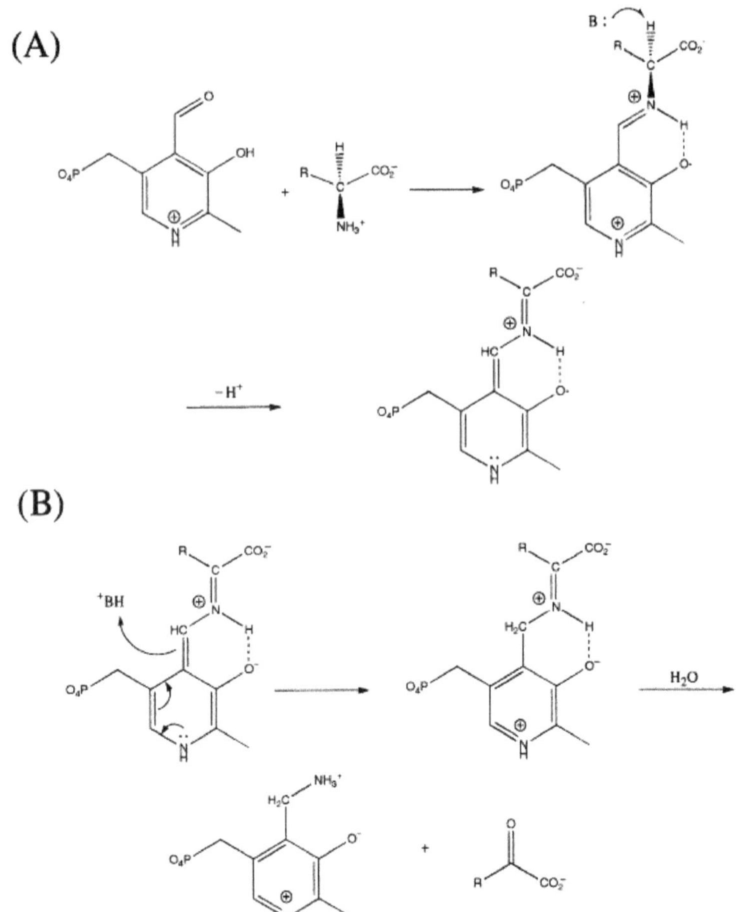

Figure 6.7 Examples of reactions of amino acids facilitated by electrophilic catalysis by pyridoxal phosphate: (A) α-hydrogen removal and (B) subsequent transamination. See text for further details.

Examples of some enzymatic reactions involving electrophilic catalysis are provided in Table 6.2.

6.3.3 General Acid/Base Catalysis

Just about every enzymatic reaction involves some type of proton transfer that requires acid and/or base group participation. In small molecule catalysis, and in some enzyme examples, protons (from hydronium ion H_3^+O) and hydroxide ions (OH-) act directly as the acid and base groups in activities referred to as *specific acid* and *specific base catalysis.* Most often, however, in enzymatic reactions organic substrates, cofactors, or amino acid side chains from the enzyme fulfill this role by acting as Brønsted–Lowry acids and bases in what is referred to as *general acid* and *general base catalysis.*

Table 6.2 Some examples of electrophilic catalysis in enzymatic Reactions

Enzyme	Electrophile
Acetoacetate decarboxylase	Lysine–substrate Schiff base
Aldolase	Lysine–substrate Schiff base
Aspartate aminotransferase	Pyridoxal phosphate
Carbonic anhydrase	Zn^{2+}
L-Malic enzyme	Mn^{2+}
Pyruvate decarboxylase	Thiamine pyrophosphate

For catalysis by small molecules (nonenzymatic reactions), general acid/ base catalysis can be distinguished from specific acid/base catalysis on the basis of the effects of acid or base concentration on reaction rate. In general acid/base catalysis, the reaction rate is dependent on the concentration of the general acid or base catalyst. Specific acid/base catalysis, in contrast, is independent of the concentrations of these species (Figure 6.8). Although most enzymatic reactions rely on general acid/ base catalysis, it is difficult to define the extent of this reliance by changing acid/base group concentration as in Figure 6.8, since the acid and base groups reside within the enzyme molecule itself. Identification of the group(s) participating in general acid/base catalysis in enzymes has generally come from studies of reaction rate pH profiles , amino acid–specific chemical modification , sitedirected mutagenesis, and x-ray crystal structures.

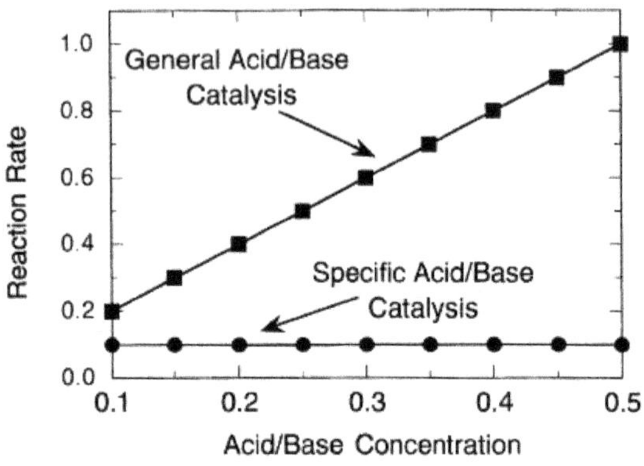

Figure 6.8 Effect of general acid or base concentration on the rate of reaction for general and specific acid- or base-catalyzed reactions.

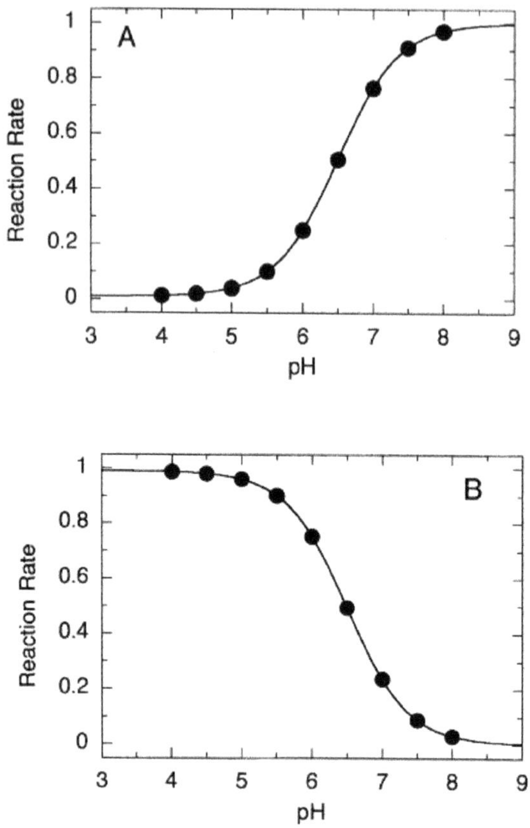

Figure 6.9 Effect of pH on the rate of reaction for (A) a general base-catalyzed reaction and (B) a general acid-catalyzed reaction.

General acids and bases will be functional only below or above their pK_a values, respectively. Hence a plot of reaction rate as a function of pH will display the type of sigmoidal curve we are used to seeing for acid—base titrations (Figure 6.9). If we plot the same data as log (reaction rate constant) as a function of pH over a finite pH range (pH 5—7 in Figure 6.9), we find a linear relationship between log (*k*) and pH. This empirical relationship is defined by the Brønsted equations for general acid and general base catalysis (Equations 6.9 and 6.10, respectively):

$$\log(k) = A - \alpha pK_a \tag{6.9}$$

$$\log(k) = A + \beta pK_a \tag{6.10}$$

where *A* is a constant associated with a specific reaction, and α and β are measures of the sensitivity of reaction to the acid or conjugate base, respectively.

The Brønsted equation for general base catalysis is similar to that described for nucleophilic catalysis. In contrast to nucleophilic catalysis, however, general base catalysis depends only on the pK_a of the catalyst and is essentially independent of the nature of the catalytic group.

Generally, these Brønsted equations indicate that the stronger the acid, the better a general acid catalyst it will be, and likewise, the stronger the base, the better a general base catalyst it will be. It is important to reemphasize, however, that the efficiency of general acid or base catalysis depends on the effective concentration of acid or base species present. The concentrations of these species depend on the pK_a of the catalyst in relation to the solution pH at which the reaction is run. For example, upon consideration of the Brønsted equations, one would say that an acid of pK_a 5 would be a better general acid than one of pK_a 7. However, if the reaction is run near pH 7, only about 1% of the stronger acid is in its acid form, while 99% of it is present as the conjugate base form. On the other hand, at the same pH, 50% of the weaker acid (pK_a 7) is present in its acid form. Thus because of the effective concentrations of the catalytically relevant forms of the two species, the weaker catalyst may be more effective at pH 7. For this reason, one finds that the reaction rates for general acid/base catalysis are maximal when the solution pH is close to the pK_a of the catalytic group. Hence, in enzymatic reactions, the general acid/base functionalities that are utilized are those with pK_a values near physiological pH (pH 7.4). Generally, this means that enzymes are restricted to using amino acid side chains with pK_a values between 4 and 10 as general acids and bases. Surveying the pK_a values of the amino acid side chains , one finds that the side chains of apartate, glutamate, histidine, cysteine, tyrosine, and lysine, along with the free N- and C-termini of the protein, are the most likely candidates for general acid/base catalysts. However, it is important to realize that the pK_a value of an amino acid side chain can be greatly perturbed by the local protein environment in which it is found. Two extreme examples of this are histidine 159 of papain, which has a side chain pK_a of 3.4, and glutamic acid 35 of lysozyme, with a greatly elevated side chain pK_a of 6.5.

The fundamental feature of general acid/base catalysis is that the catalytic group participates in proton tranfers that stabilize the transition state of the chemical reaction. A good example of this comes from the hydrolysis of ester bonds in water, a reaction carried out by many hydrolytic enzymes. The mechanism of ester hydrolysis requires formation of a transition state involving partial charge transfer between the ester and a water molecule, as shown in Figure 6.10. This transition state can be stabilized by a basic group (B:) acting as a partial proton acceptor from the water molecule, thus enhancing the stability of the partial positive charge on the water molecule (Figure 6.10B).

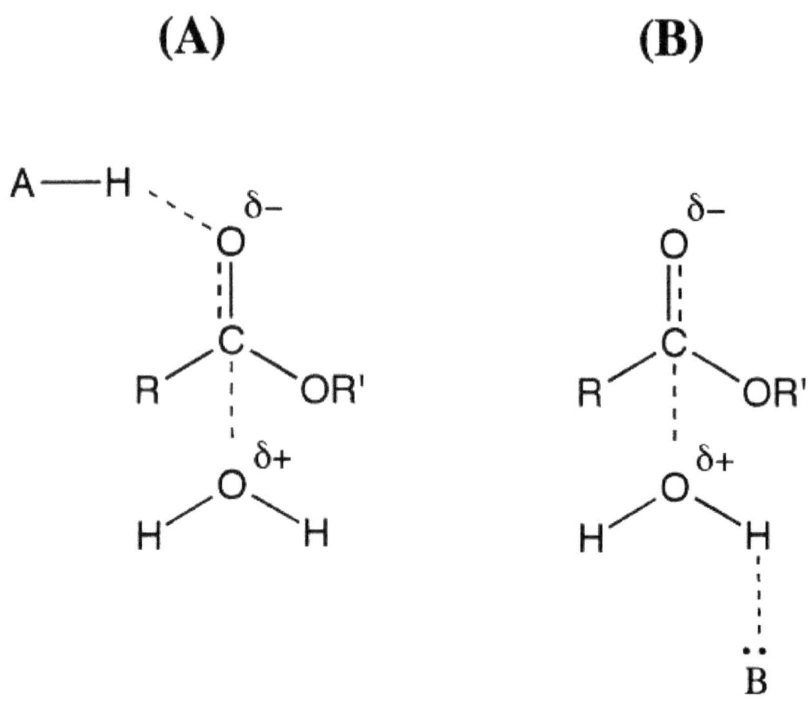

Figure 6.10 Transition state stabilization by a general acid (A) or general base (B) in ester bond hydrolysis by water.

Alternatively, the transition state can be stabilized by an acidic group that acts as a partial proton donor to the carbonyl oxygen of the ester (Figure 6.10A).

This type of acid/base group transition state stabilization is very common in enzyme catalysis. The active site histidine residue of the serine proteases, for example, acts in this capacity to stabilize the transition state during peptide hydrolysis, as described later.

As we have just stated, general acid/base catalysis is a common mechanism of transition state stabilization in enzymatic catalysis. So too is nucleophilic catalysis, as described in Section 6.3.2. Ambiguity can arise in distinguishing nucleophilic catalysis from general base catalysis because these two mechanisms share some common kinetic features. Bender et al. suggest the following criteria for distinguishing betweeen these mechanisms:

1. 1. If the leaving group of a reaction can itself catalyze the same reaction, this is evidence of general base rather than nucleophilic catalysis. Consider the reaction:

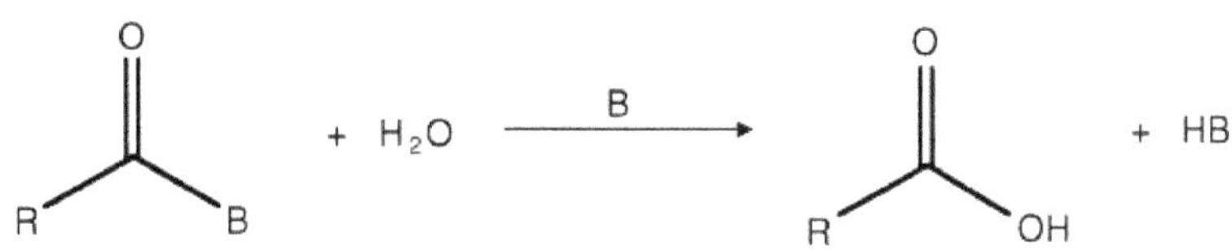

If this reaction went with nucleophilic attack of B on the carbonyl, the reaction product would be the same as the starting material. Hence, catalysis by B indicates this species must be functioning as a general base.

2. Catalysis by a second equivalent of an attacking species is evidence that general base catalysis is occurring. The covalent intermediate of nucleophilic catalysis forms stoichiometrically with substrate. Hence catalysis by a second equivalent of the attacking species cannot occur via nucleophilic catalysis,
3. As stated above, observation of a transient intermediate that can be identified as a covalent adduct constitutes proof of nucleophilic catalysis, rather than the general base version.
4. As stated above, the Brønsted plot for general base catalysis fits a single straight line for bases of identical basicity, regardless of their structural type. Conversely, nucleophilic catalysis is characterized by Brønsted plots with large differences in rates when one is considering, for example, nitrogen and oxygen pairs of catalysts of equal basicity.
5. Related to point 4, steric hinderance is not an important determinant in general base catalysis but can have significant effects on nucleophilic catalysis. This follows because general base catalysis involves proton abstraction, while nucleophilic catalysis involves attack at a carbon center; the steric requirements for the latter reaction are far greater than that of the former.

6. Deceleration of reaction rate by addition of a common ion is an indication of a reaction mechanism involving a reversible intermediate, hence nucleophilic catalysis. For example, the pyridine-catalyzed hydrolysis of acetic anhydride proceeds with formation of acetateion:

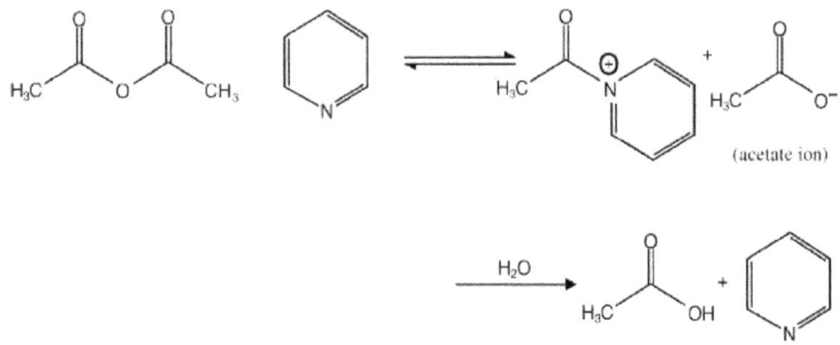

Addition of acetate ion significantly decreases reaction rate, indicating nucleophilic catalysis by pyridine.

Additional criteria that are more relevant to small molecule catalysis are provided by Bender et al. . The interested reader should consult this text and the references therein.

6.3.4 Conformational Distortion

Two experimental observations lead to the hypothesis that conformational adjustments of the enzyme, the substrate, or both are important aspects of enzymatic catalysis.

First, quantitative studies of enzymatic reactions often lead to the conclusion that the observed rate enhancement cannot be accounted for fully by consideration of approximation, covalent catalysis, and acid/base catalysis alone. Hence, an additional mechanism for rate enhancement must be invoked. It is well established that many proteins, including enzymes, are conformation ally dynamic. Therefore, one hypothesis that has emerged is that conformational adjustments are used to distort the substrate, the enzyme, or both to bring the system toward the transition state structure. Figure 6.11 illustrates this concept in schematic form for a bond cleavage reaction. In this example the enzyme active site undergoes a series of conformational adjustments that distort the substrate toward its transition state structure, thus facilitating catalytic bond rupture. In this illustration, the distortion of the substrate by the enzyme active site is akin to the medieval torture in which a prisioner was stretched on a device known as a rack. For this reason, the term *rack mechanism* is sometimes used to describe the induced-strain hypothesis .

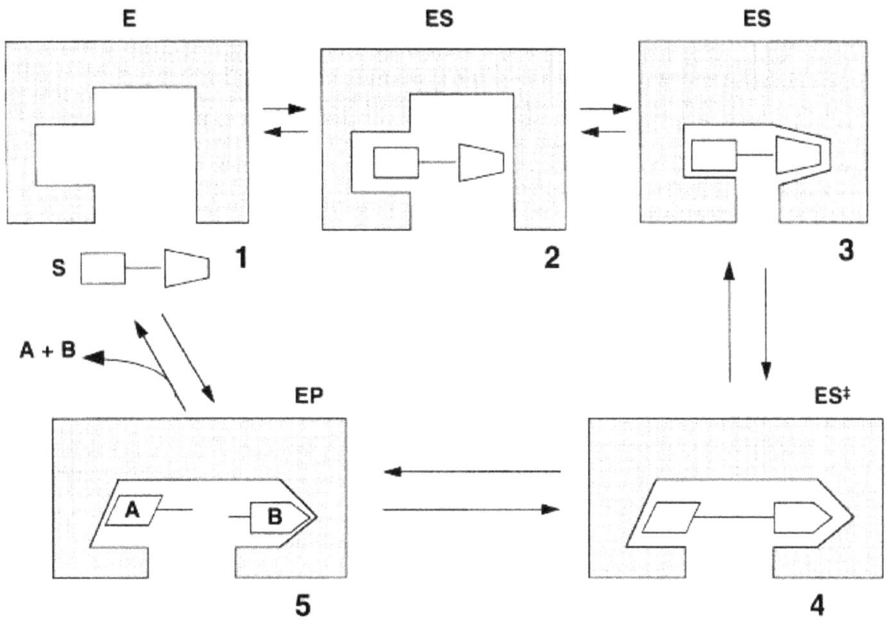

Figure 6.11 The rack model of enzyme–substrate interaction and catalysis: 1, the free enzyme is in solution with the free substrate; 2, initial binding occurs to form the ES complex without structural distortion of the substrate; 3, the enzyme undergoes a conformational change to induce better binding with the substrate (induced fit); 4, the enzyme undergoes further conformational adjustments that induce strain into the substrate molecule, distorting it from its ground state structure toward its transition state structure; 5, the strain induced in 4 leads to bond rupture and release of the two hydrolytic products, A and B.

The second experimental observation that suggests a need to invoke conformational distortions is the manifestation of substrate specificity in k_m, rather than in K_{cat}. We saw in Chapter 5 that the best measure of substrate specificity is the second-order rate constant obtained by dividing k_{cat} by K_m. If substrate specificity merely involved discrimination in binding of substrates in their ground states, one would expect that differences in K_m would be the dominant factor in distinguishing a good from a bad substrate. Experimentally, however, one often observes that substrate specificity is manifested at high, saturating substrate concentration, rather than at low substrate concentration.

In other words, specificity is largely dictated by differences in k_{cat} (i.e., V_{max}), rather than by differences in K_m. A good example comes from studies of the hydrolysis of synthetic peptides by the enzyme pepsin ; the results of these studies are summarized in Table 6.3, which indicates that for a series of peptide substrates, the catalytic efficiency (measured as k_{cat}/K_m) of pepsin varies over 1000-fold. Looking at the individual kinetic constants, we see that k_{cat} likewise varies over about 1000-fold. In contrast, the values for K_m of the various substrates are quite similar, differing only about four-fold at most. In fact, the K_m value for the worst substrate is actually lower than that of the best substrate, clearly indicating that ground state substrate binding is not a major determinant of substrate specificity for this enzyme. Instead, the experimental data for pepsin, and for a large number of other enzymes, suggest that substrate specificity is determined mainly by the transition states structure, rather than by that of the ground state.

Table 6.3 Steady state kinetics of synthetic peptide hydrolysis by Pepsin

Peptide[a]	K_m(mM)	k_{cat}(s^{-1})	k_{cat}/K_m ($mM^{-1}\cdot s^{-1}$)
Cbz-G-H-F-F-OEt	0.8	2.4300	3.04000
Cbz-H-F-W-OEt	0.2	0.5100	2.55000
Cbz-H-F-F-OEt	0.2	0.3100	1.55000
Cbz-H-F-Y-OEt	0.2	0.1600	0.80000
Cbz-H-Y-F-OEt	0.7	0.0130	0.01860
Cbz-H-Y-Y-OEt	0.2	0.0094	0.04700
Cbz-H-F-L-OMe	0.6	0.0025	0.00417

[a]Cbz, carbobenzyloxy; OEt, ethyl ester of carboxy terminus; OMe, methyl ester of carboxy terminus.

A similar conclusion is often reached by the study of the effects of point mutations within the active site of enzymes. For example, Leatherbarrow et al. studied the enzyme tyrosyl—tRNA synthetase, an enzyme that covalently links the amino acid tyrosine to its tRNA. The enzyme requires ATP for catalysis and proceeds through formation of an enzyme-bound tyrosyl—ATP transition state.

The crystal structure of the enzyme with the covalent intermediate tyrosyl—AMP bond was known, and Leatherbarrow et al. were able to use this to model the transition state complex. From this modeling the workers identified two amino acid residues within the enzyme active site that should be critical for catalysis: Thr 40 and His 45. They then set about making three mutant enzymes: a Thr 40-Ala mutant, a His 45-Gly mutant, and the double mutant Thr 40-Ala/His 45-Gly. When they compared the kinetic parameters for these three mutants to the wild-type enzyme, they found that the values of K_S for either tyrosine or ATP were hardly affected; the largest change in K_S seen was about four-fold relative to the wild-type enzyme. In contrast, however, the values of k_{cat} were dramatically changed. The value of k_{cat} went from 38 s^{-1} for the wild-type enzyme to 0.16 s^{-1} for the His 45-Gly mutant, to 0.0055 s^{-1} for the

Thr 40-Ala mutant, and to 0.00012 s^{-1} for the double mutant. Thus Leatherbarrow et al. concluded that the main function of these residues was to provide favorable binding interactions with the tyrosyl—ATP transition state structure, and in contrast provided no binding interactions with the ground state substrate.

Jencks has argued that for reversible enzymatic reactions, distortion of the ES complex is a necessary component of catalysis. Consider a reversible enzymatic reaction that proceeds through nucleophilic attack of a group on the substrate molecule. We could imagine that the enzyme active site and the substrate ground state are both conformationally rigid. The substrate binds to the active site because its ground state configuration fits perfectly into this pocket. In this case, the interatomic distance between the enzymatic nucleophile and the substrate carbon atom that is attacked would be determined by the van der Waals radii of the interacting atoms. If this were so, the rigid product molecule could not have a perfect fit to the same rigid active site structure. Thus the enzyme—product complex would be destabilized relative to the enzyme—substrate complex as a result of steric hindrance (hence loss of binding energy).

The enzyme could overcome this by distorting the bound product in such a way as to more closely resemble the substrate molecule. This would facilitate reaction in the reverse direction. Likewise, if the active site were rigid and fit the product molecule perfectly, the imperfect fit of the substrate molecule would now impede reaction in the forward direction. The need for enzymatic rate acceleration in both the forward and reverse reactions is most effectively accomplished by having the active site structure best matched to a structure intermediate between the substrate and product states—that is, by having the active site designed to best match the transition state structure.

However, the foregoing arguments and experimental data are not adequately accounted for by models in which the ground state substrate and the enzyme active site are conformationally rigid, as in the original lock and key model of Fischer. Hence, models are needed that take into account the need for conformational flexibility and optimization of active site—transition state complementarity. Three major models have been put forth to fill these needs: (1) the induced-fit model, (2) the nonproductive binding model, and (3) the induced-strain model.

The induced-fit model, first proposed by Koshland , suggests that the enzyme active site is conformationally fluid. In the absence of substrate, the active site is in a conformation that does not support catalysis. When a "good" substrate binds to the active site, the binding forces between the enzyme and the substrate are used to drive the enzyme into an energetically less favorable, but catalytically active conformation (the rack model illustrated in Figure 6.11 is one interpretation of this induced fit model). In this model a "poor" substrate lacks the necessary structural features to induce the conformational change required for catalytic activity, and thus does not undergo reaction. The expected results from the induced-fit model is that V_{max} is optimal for specific substrates that can best utilize binding interactions to induce the necessary conformational activation of the enzyme molecule.

In the nonproductive binding model the enzyme active site is considered to be rigid; and a "good" substrate would be one that has several structural features, each one being complementary to a specific subsite within the active site structure. Because of the presence of multiple complementary subsite interactions, there would be only one active conformation and orientation of the substrate in the active site of the enzyme; other structures or orientations would lead to binding of an inactive substrate species that would be nonproductive with respect to catalysis. A "poor" substrate in this model might lack one or more key functional group for correct binding. Alternatively, the functional group might be present in a "poor" substrate, but arranged in a fashion that is inappropriate for correct binding in the enzyme active site; these concepts are illustrated in Figure 6.12. The expected result from this model is that V_{max} for a "poor" substrate would be greatly diminished because only a fraction of the "poor" substrate—that happened to bind in a productive mode—would lead to catalytic activity. As discussed , the nonproductive binding model is often invoked to explain the phenomenon of substrate inhibition at high substrate concentrations.

The third major model for conformational distortion is referred to as the induced-strain model. In this model the binding forces between enzyme and substrate are directly used to induce strain in the substrate molecule, distorting it toward the transition state structure to facilitate reaction. The enzyme active site is considered to be flexible in this model. The most stable (i.e., lowest energy) conformation of the active site is one that does not optimally fit the ground state substrate conformation, but instead is complementary to the transition state of the reaction. For the ground state substrate to bind, the enzyme must undergo a conformational deformation that is energetically unfavorable. Hence, in the ES complex there will be a driving force for the enzyme molecule to return to its original lower energy conformation, and this will be accompanied by distortions of the substrate molecule to bring it into the transition state structure that is complementary to the lowest energy conformation of the active site (Figure 6.13). At first glance, this model seems similar to the induced-fit model, in which binding energy is used to bring reactive groups within the active site into proper register with respect to a "good" substrate to provide substrate specificity. In contrast, the induced-strain model utilizes the binding energy to directly decrease the free energy of activation for the reaction by coupling substrate distortion to an energetically favorable conformational relaxation of the enzyme molecule. Steric effects are not the only mechanism for inducing strain in a bound substrate molecule. Electrostatic effects can also provide for induction of strain in the ground state, which is subsequently relieved in the transition state structure. For example, a negatively charged amino acid side chain in the hydrophobic enzyme active site would have a strong polarizing effect that would be unfavorable to binding of an uncharged ground state substrate. If, however, the transition state involved formation of a cationic center, the negative counterion would now provide a favorable interaction. This relief of ground state destabilization can be effectively used as a driving force for transition state formation. Likewise, hydrogen bonding, hydrophobic interactions, and other noncovalent forces can provide unfavorable interactions with the ground state substrate that drive distortion to the transition state structure, where these interactions are energetically favorable.

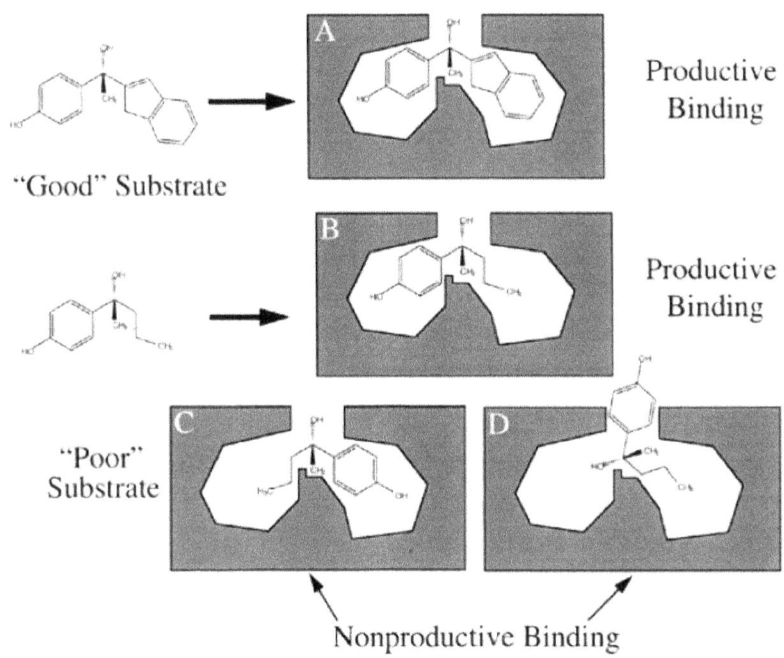

Figure 6.12 Schematic illustration of the nonproductive binding model. In this model a "good" substrate has specific groups that engage subsites in the enzyme active site, resulting in a unique, productive binding orientation that facilitates catalysis. A "poor" substrate is one that lacks a keep functionality to help orient the substrate properly in the active site. The poor substrate therefore may bind in a number of orientations, only a fraction of which lead to efficient catalysis. The other, nonproductive, binding orientations are inhibitory to catalysis.

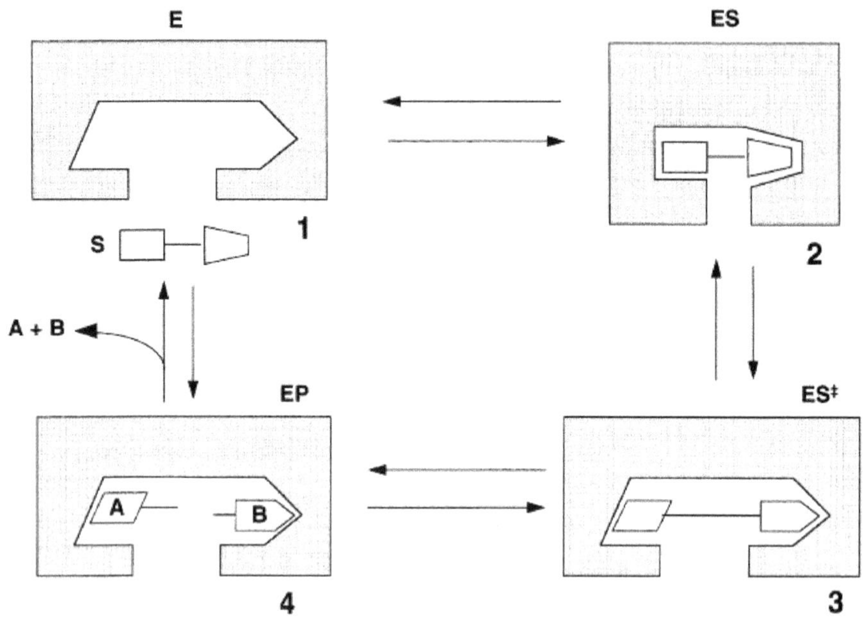

Figure 6.13 Schematic illustration of the induced-strain model: the free enzyme (1) is in its lowest energy conformation that is most complementary to the transition state structure of the substrate. Upon substrate binding (2), the enzyme undergoes a conformational transition to engage the substrate. The conformational strain induced in (2) is relieved by formation of the bound transition state in 3. Catalysis then leads to formation of the enzyme–product complex (4), before product release to re-form the free enzyme.

6.3.5 Preorganized Active Site Complementarity to the Transition State

An alternative to mechanisms utilizing conformational distortion is one in which the enzyme active site is, relatively speaking, conformationally rigid and preorganized to optimally fit the substrate in its transition state conformation. This is somewhat reminiscent of the lock and key model of Fischer, but here the complementarity is with the substrate transition state, rather than the ground state. In a recent review of this mechanism, Cannon and Benkovic (1998) use the following thermodynamic cycle to define an equilibrium dissociation constant for the ES‡ binary complex, K_{TS}, which is equivalent to the ratio $(k_{cat}/K_m)/k_{non}$, where k_{non} is the rate constant for formation of the substrate transition state in the absence of enzyme catalysis.

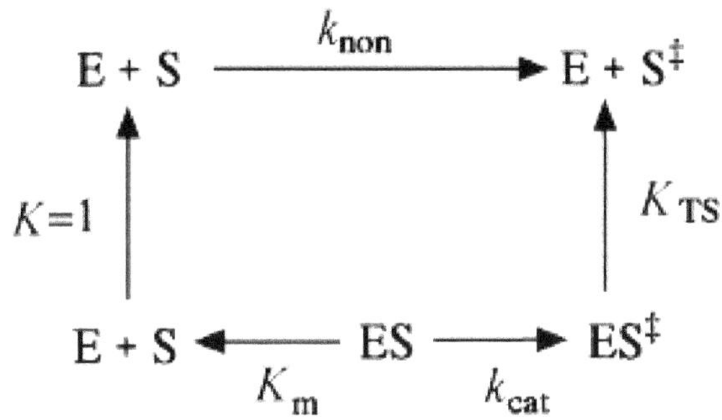

From our earlier discussion, in which we learned that enzymes bind the transition state better than the ground state substrate, it is clear that K_{TS} is typically much smaller than K_S. Hence there is usually a large, favorable free energy of binding for the ES‡ complex. Cannon and Benkovic suggest two potential ways to explain the difference in free energy of binding between S‡ and S. First, there may be significantly stronger binding interactions between the enzyme and the transition state conformation of the substrate. This possibility was discussed above, and, at least for some enzymes there is experimental evidence for stronger interaction between the enzyme and transition state analogues, than between the enzyme and the ground state substrate.

The second possibility is that the free energy of interaction between the solvent and the transition state is very much less favorable than that for the solvent and the ground state substrate. Hence, in solution the attainment of the transition state must overcome a significant free energy change due to solvation effects. In other words, in this case significantly better interactions between the enzyme and S‡ relative to S are not the main driving forces for transition state stabilization.

Instead, by removing S from solvent, the enzyme avoids much of the solvation-related barrier to formation of S‡. In this model the enzyme does not so much stabilize the transition state as avoid the destabilizing effect of the solvent by sequestering the substrate. Cannon and Benkovic suggest that both possibilities occur in enzymatic catalysis, but that the latter is a more dominant effect.

As evidence for this mechanism, Cannon and Benkovic have plotted K_{TS} against k_{cat} and k_{non} for a series of first-order, or pseudo-first-order, enzymatic reactions and appropriate model reactions in solution. If the main function of enzymes were to bind S‡ much more tightly than S, one would expect that the catalytically most powerful enzymes (i.e., those with the lowest values of K_{TS}) would display the fastest turnover rates (k_{cat}). On the other hand, if rate acceleration by enzymes is due mainly to relief of the retardation of solution reactions due to solvation effects, the enzymes displaying small K_{TS} values would be those catalyzing reactions that proceed very slowly in solution (i.e., with small values of k_{non}). The data plot provided by the authors shows a significant, negative correlation between the values of K_{TS} and k_{non}, but a general lack of correlation between K_{TS} and k_{cat}. Cannon and Benkovic conclude from this plot that the kinetics of reactions in solution are the dominant determinants of K_{TS}.

The physical explanation for these observations is that the effect of solvent on the solution reactions is counterproductive. Significant solvent reorganization is required for the solution reaction to proceed to the transition state, and this reorganization has a retarding effect on the rate of reaction. The enzyme thus functions as a mechanism for solvent substitution for the reactants.

Cannon and Benkovic point out that the solvent effect on reaction will depend on both the dielectric response and the polarity of the medium. Because enzyme active sites are largely hydrophobic, they will generally have low dielectric constants; but they can be highly polar (as a result of the effect of placing a charge within a low dielectric medium), thus producing very intense electric fields. By judicious placement of charged groups within the active site, the enzyme can achieve electrostatic complementarity with the transition state structure of the substrate, thus eliminating the solvent retardation effects.

The foregoing argument suggests that the enzyme active site is preorganized to be complementary to the transition state of the substrate, thus minimizing the energetic cost of reorganizations such as those occurring in solution. This mechanism would disfavor conformationally induced distortions of the substrate, which would only add to the reorganizational cost to catalysis. Cannon and Benkovic point out further that the time scale of most protein conformational changes is inconsistent with the rates of catalysis for many enzymes. It is important to note, however, that vibrational bond motions occur on a time scale (ca. 10^{-14}—10^{-13} s^{-1}) that is consistent with enzymatic catalysis. Hence, small vibrational adjustments of the enzyme—substrate complex cannot be ruled out by the foregoing argument.

Thus, the mechanism proposed by Cannon and Benkovic relies on the preorganization of the enzyme active site in a configuration that is complementary to the transition state of the substrate (but does not necessarily bind the transition state extremely tightly). This preorganized active site features strategically located reactive groups (general acids/bases; active site nucleophiles, hydrogen-bonding partners, etc.) in a low dielectric medium that greatly relieves the destabilization of the transition state associated with reaction in solution. Thus, in this mechanism, enzymes primarily accelerate reaction rates not by stabilizing the transition state through tight binding interactions per se, but instead by avoiding the energetic penalties that accompany transition state formation in bulk solution. The authors point out that this mechanism suggests that one could capture most of the catalytic efficiency of enzymes by providing a properly preorganized transition state binding pocket within an engineered protein. For example, the immune system normally produces antibodies that recognize and bind tightly to proteins and peptides from an infecting organism.

Antibodies can also be produced that recognize and bind small molecules, referred to as haptens. One might therefore expect antibodies raised against transition state analogues of specific reactions to display some ability to act as catalysis of the reaction. This approach has been experimentally verified , although the resulting *catalytic antibodies* have thus far showed only modest catalytic activity relative to natural enzymes.

6.4 THE SERINE PROTEASES: AN ILLUSTRATIVE EXAMPLE

The serine proteases are a family of enzymes that catalyze the cleavage of specific peptide bonds in proteins and peptides. As we briefly mentioned earlier , the serine proteases have a common mechanism of catalysis that requires a triad of amino acids at the active sites of these enzymes; a serine residue (hence the family name) acts as the primary nucleophile for attack of the peptide bond, and the nucleophilicity of this group is enhanced by specific interactions with a histidine side chain (a general acid/base catalyst), which in turn interacts with an aspartate side chain. The catalytic importance of the active site serine and histidine residues has been demonstrated recently by site-directed mutagenesis studies, in which replacement of either the serine or the histidine or both reduced the rate of reaction by the enzyme by as much as 10^6 fold .

Because of their ease of isolation and availability in large quantities from the gastric juices of large animals, the serine proteases were among the first enzymes studied. While the members of this family studied initially were all digestive enzymes, we now know that the serine proteases perform a wide variety of catalytic functions in most organisms from bacteria to higher mammals.

In man, for example, serine proteases take part not only in digestiveprocesses, but also in the blood clotting cascade, inflammation, wound healing, general immune response, and other physiologically important events. These enzymes are among the most well-studied proteins in biochemistry. A great deal of structural and mechanistic information on this class of enzymes is available from crystallographic, classical biochemical, mechanistic, and mutagenesis studies . Because of this wealth of information, the serine proteases provide a model for discussing in concrete chemical and structural terms some of the concepts of substrate binding specificity and transition state stabilization covered thus far in this chapter.

To cleave a peptide bond within a polypeptide or protein, a protease must recognize and bind a region of the polypeptide chain that brackets the scissile peptide bond (i.e., the bond that is to be cleaved). Proteases vary in the length of polypeptide that forms their respective recognition sequences, but most bind several amino acid residues in their active sites. A nomenclature system has been proposed by Schechter and Berger to keep track of the substrate amino acid residues involved in binding and catalysis, and the corresponding sites in the enzyme active site where these residues make contact. In this system, the bond that is to be hydrolyzed is formed between residue P1 and P1' of the substrate; P1 is the residue that is on the N-terminal side of the scissile bond, and P1' is the C-terminal hydrolyzed residue (the "P" stands for "peptide" to designate these residues as belonging to the substrate of the reaction). The residue adjacent to P1 on the N-terminal side of the scissile bond is designated P2, and the residue adjacent to the P1' residue on the C-terminal side is P2'. The "subsite" within the enzyme active site that residue P1 fits into is designated S1, and the "subsite" into which residue P1' fits is designated S1'.

The numbering continues in this manner, as illustrated in Figure 6.14 for a six-residue peptide substrate. We shall use this nomenclature system from now on when discussing proteolytic enzymes.

On the basis of their structural properties, the serine proteases have been divided into three classes, called the chymotrypsin-like, the subtilisin-like, andthe serine carboxypeptidase II—like families. The secondary and tertiary structures of the proteins vary considerably from one family to another, yet in all three families the active site serine, histidine, and aspartate are conserved and a common mechanism of catalysis is used. All these enzymes catalyze the hydrolysis of ester and peptide bonds through the same acyl transfer mechanism (Figure 6.15), with a rate acceleration of 10^9 or more relative to the uncatalyzed reaction. After formation of the ES complex, the carbonyl carbon of the scissile peptide bond (i.e., that on P1) is attacked by the active site serine, forming a tetrahedral intermediate with an oxyanionic center on the carbonyl carbon that is highly reminiscent of the transition state of the reaction. This transition state is stabilized by specific hydrogen-bonding interactions between residues in the active site pocket and the oxyanion center of the substrate. In subtilisin this hydrogen bonding is provided by the backbone nitrogen of Ser 195 (the active site nucleophile) and the side chain of Asn 155. In the chymotrypsin-like enzymes, these H bonds are provided by the backbone nitrogens of the nucleophilic serine residue and an active site glycine, while in the serine carboxypeptidases these bonds are formed by the backbone nitrogens of a tyrosine and a glycine in the active site.

Crystallographic data from studies of subtilisin indicate that a weak hydrogen bond exists between Asn 155 and the substrate in the ES complex, and that this H-bonding is significantly strengthened in the transition state (an example of conformational alterations in the enzyme active site that facilitate transition state stabilization).

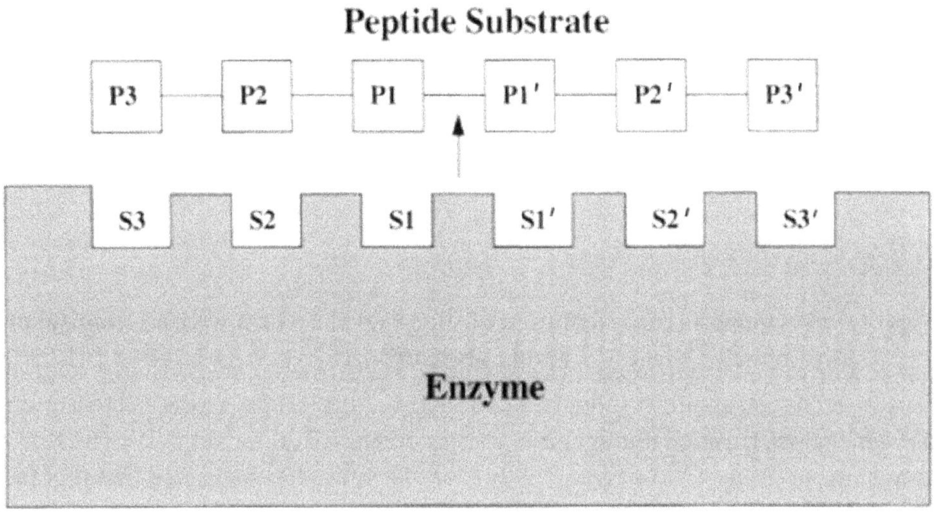

Figure 6.14 The protease subsite nomenclature of Schechter and Berger (1967): residues on the peptide substrate are labeled P1–Pn on the N-terminal side of the scissile bond, and P1'–Pn' on the C-terminal side of this bond; the scissile bond is thus between residues P1 and P1'. The corresponding subsites into which these residues fit in the enzyme active site are labeled S1–Sn and S1'–Sn'.

Mutation of Asn 155 to any of a variety of non-hydrogen-bonding amino acids significantly decreases the rate of the enzymatic reaction, as would be expected from our discussion of enzymatic rate enhancement by transition state stabilization.

The transition state then decays as a proton is donated from the active site histidine to the amine group of P1', followed by dissociation of the first product of the reaction, the peptide starting at P1', and simultaneous formation of a covalent intermediate with an acyl group (from P1) bound to the active site serine. The enzyme is then deacylated by nucleophilic attack by a water molecule that enters the enzyme active site from the cavity resulting from the departure of the first product peptide. The deacylation reaction proceeds with formation of another tetrahedral transition state, very similar to that formed during the acylation reaction, and engaging the same stabilizing H bonds with the enzyme. This transition state decays with proton transfer to the active site histidine and release of the second peptide product having the P1' group at its amino terminus.

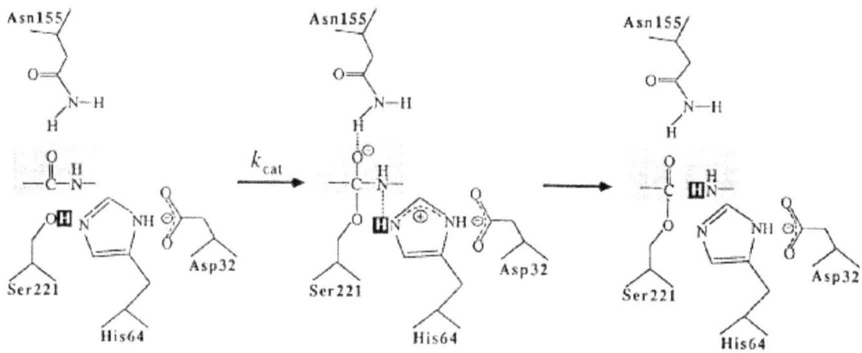

E·S E·S‡ E-Ac

Figure 6.15 Schematic representation of the general acyl transfer mechanism of serine proteases.

The active site aspartate residue is a common feature of the serine proteases and has been shown, through mutagenesis studies, to be critical for catalysis. The role of this residue in catalysis is not completely clear. Early studies suggested that together with the serine and histidine residues, this aspartate formed a catalytic "triad" that acts as a proton shuttle. Such a specific interaction requires a precise geometric relationship between the side chains of the aspartate and histidine residues to ensure strong H-bonding. However, the orientation of the aspartate side chain relative to the Ser-His active site residues varies considerably among the three classes of serine protease, making it unlikely that direct proton transfer occurs between the histidine and aspartate side chains in all these enzymes. In fact, some workers have suggested that the catalytic machinery of the serine proteases is most correctly viewed as two distinct catalytic dyads—one comprising the serine and histidine residue and the other comprising the histidine and aspartate residues—rather than as a single catalytic triad . Regardless of the molecular details, experimental data demonstrate that the presence of the carboxylate anion of the aspartate influences the reactivity of the histidine residues in a way that is critical for catalysis.

The foregoing discussion provides a good example of the interplay between substrate and enzyme active site that must accompany transition state stabilization and reaction rate enhancement. Substrate binding, however, must precede these catalytic steps, and formation of the enzyme—substrate complex is also governed by the stereochemical relationships between groups on the substrate and their counterpart subsites in the enzyme active site. For example, the differences in substrate specificity within the chymotrypsin-like and subtilisin- like serine proteases can be explained on the basis of their active site structures. In all these enzymes, substrate binding is facilitated by H-bonding to form β-pleated sheet structures between residues in the enzyme active site and the P1—P4 residues of the substrate (Figure 6.16). These interactions provide binding affinity for the substrates but do not significantly differentiate one peptide substrate from another. The P1—S1 site interactions appear to play a major role in defining substrate specificity in these enzymes. Subtilisins generally show broad substrate specificity, with a preference for large, hydrophobic groups at the substrate P1 position. The order of preference is approximately Tyr, Phe>Leu, Met, Lys>His, Ala, Gln, Ser>Glu, Gly . The S1 site of the subtilisins exists as a broad, shallow cleft that is formed by two strands of β-sheet structure and a loop region of variable size (Figure 6.16B). In a subclass of these enzymes, in which a P1 Lys residue is accommodated, the loop contains a glutamate residue positioned to form a salt bridge with the substrate lysine residue,to neutralize the charge upon substrate binding.

In the chymotrypsin-like enzymes, the S1 pocket consists of a deep cleft into which the substrate P1 residue must fit (Figure 6.16A). The identity of the amino acid residues within this cleft will influence the types of substrate residue that are tolerated at this site. The chymotrypsin-like enzymes can be further subdivided into three subclasses on this basis. Enzymes within the trypsinlike subclass have a conserved aspartate residue at position 189, located at the bottom of the S1 well; this explains the high preference of these enzymes for substrates with arginine or lysine residues at P1. This aspartate is replaced by a serine or small hydrophobic residue in the chymotrypsin and elastase subclasses; hence both subclasses show specificity for nonpolar P1 residues. The other amino acid residues lining the S1 pocket further influence substrate specificity. The elastase subclass contains in this pocket large nonpolar groups that tend to exclude bulky substrate residues.

Hence, the elastase subclass favors substrates with small hydrophobic residues at P1. In contrast, the chymotrypsin subclass has small residues in these positions (e.g., glycines), and these enzymes thus favor larger hydrophobic residues, such as tyrosine and phenylalanine at the P1 position of their substrates.

Within this overview of the serine proteases we have observed specific examples of how the active site structure of enzymes (1) engages the substrate and binds it in an appropriate orientation for catalysis (e.g., the H-bonding network developed between the P1—P4 residues of the substrate and the active site residues), (2) stabilizes the transition state to accelerate the reaction rate (e.g., the stabilization of the tetrahedral oxyanionic intermediate through H-bonding interactions), and (3) differentiates between potential substrates on the basis of stereochemical relationships between the substrate and active site subsites. While the molecular details differ from one enzyme to another, the general types of interaction illustrated with the serine proteases also govern substrate binding and chemical transformations in all the enzymes nature has devised.

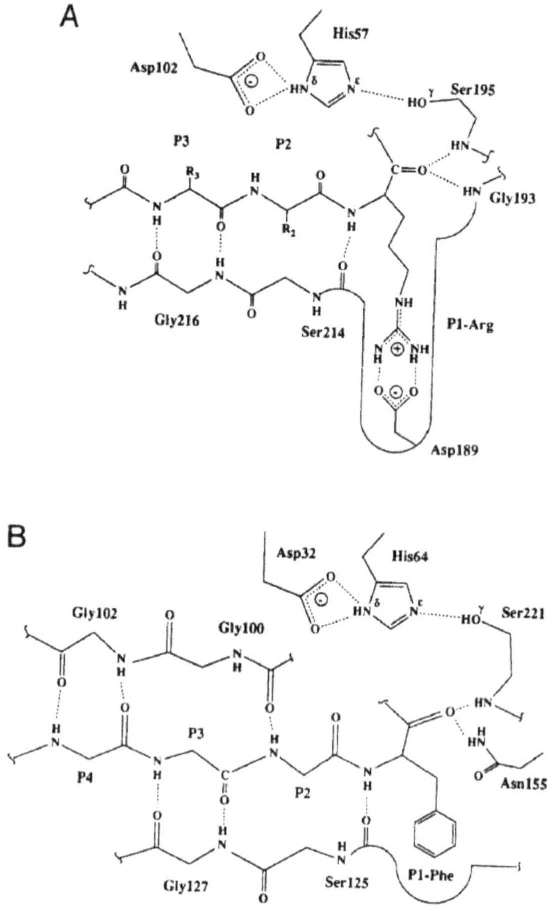

Figure 6.16 Substrate—active site interactions in the serine proteases. (A) Interactions within the trypsinlike class of serine proteases. (B) Interactions within the subtilisin-like class of serine proteases.

6.5 ENZYMATIC REACTION NOMENCLATURE

The hydrolytic activity illustrated by the serine proteases is but one of a wide variety of bond cleavage and bond formation reactions catalyzed by enzymes. From the earliest studies of these proteins, scientists have attempted to categorize them by the nature of the reactions they provide. Group names have been assigned to enzymes that share common reactivities. For example, "protease" and "proteinase" are used to collectively refer to enzymes that hydrolyze peptide bonds. Common names for particular enzymes are not always universally used, however, and their application in individual cases can lead to confusion. For example, there is a metalloproteinase known by the common names stromelysin, MMP-3 (for matrix metalloproteinase number 3), transin, and proteoglycanase. Some workers refer to this enzyme as stromelysin, others call it MMP-3, and still others call it transin or proteoglycanase.

A newcomer to the metalloproteinase field could be quite frustrated by this confusing nomenclature. For this reason, the International Union of Pure and Applied Chemistry (IUPAC) formed the Enzyme Commission (EC) to develop a systematic numerical nomenclature for enzymes. While most workers still use common names for the enzymes they are working with, literature references should always include the IUPAC EC designations, which have been universally accepted, to let the reader know precisely what enzymes are being discussed. The EC classifications are based on the reactions that enzymes catalyze. Six general categories have been defined, as summarized in Table 6.4. Within each of these broad categories, the enzymes are further differentiated by a second number that more specifically defines the substrates on which they act. For example, 11 types of hydrolase (category 3) can be defined, as summarized in Table 6.5.

Table 6.4 The IUPAC EC classification of enzymes into six general categories according to the reactions they catalyze

First EC Number	Enzyme Class	Reaction
1	Oxidoreductases	Oxidation–reduction
2	Transferases	Chemical group transfers
3	Hydrolases	Hydrolytic bond cleavages
4	Lyases	Nonhydrolytic bond cleavages
5	Isomerases	Changes in arrangements of atoms in molecules
6	Ligases	Joining together of two or more molecules

Individual enzymes in each subclass are further defined by a third and a fourth number. In this way any particular enzyme can be uniquely identified.Examples of the common names for some enzymes, and their EC designations are given in Table 6.6.

Table 6.5 The IUPAC EC subclassifications of the hydrolases

First Two EC Numbers	Substrates[a]
3.1	Esters, —C(O)—O- - - R, or with S or P replacing C, or —C(O)—S- - - R
3.2	Glycosyl, sugar—C—O- - - R, or with N or S replacing O
3.3	Ether, R—O- - - R, or with S replacing O
3.4	Peptides, C- - - N
3.5	Nonpeptides C- - - N
3.6	Acid anhydrides, R—C(O)—O- - - C(O)—R′
3.7	C- - - C
3.8	Halides (X), C- - - X, or with P replacing C
3.9	P- - - N
3.10	S- - - N
3.11	C- - - P

[a]Hydrolyzed bonds shown as dashed lines.

Table 6.6 Some examples of enzyme common names and their EC designations

Common Name(s)	EC Designation
Cytochrome oxidases (cytochrome *c* oxidase)	EC 1.9.3.1
Prostaglandin G/H synthase (cyclooxygenase)	EC 1.14.99.1
Stromelysin (MMP-3, proteoglycanase)	EC 3.4.24.17
Dihydroorotate dehydrogenase	ED 1.3.99.11
Rhodopsin kinase	EC 2.7.1.125

MD). This volume lists the common names of chemicals and reagents (including enzymes) that are referred to in the medical literature covered by the *Index Medicus* (a source book for literature searching of medically related subjects). Enzymes are listed here under their common names (with cross-references for enzymes having more than one common name) and the EC designation is provided for each. Most college and university libraries carry the *Index Medicus* and will have this supplement available, or one can purchase the supplement directly from the National Library of Medicine. Yet another resource for determining the EC designation of an enzyme is the Enzyme Data Bank, which can be accessed on the Internet.* This data bank provides EC numbers, recommended names, alternative names, catalytic activities, information on cofactor utilization, and associated diseases for a very large collection of enzymes..

Chapter 7

PH STUDIES WITH ENZYMES

Enzyme-catalyzed reactions involve one or more proton transfers. Acid–base chemistry permeates most of enzyme chemical mechanisms. Ionizable amino acid side chains of the enzyme protein are typically involved in such catalysis. Each ionizable group can be viewed as an acid and also a conjugate base.

Interactions of enzyme with its substrate (or any ligand for that matter) involve at least one or few groups whose correct ionization is necessary for its optimal function. The ionization state of acid–base groups on the enzyme, substrate, and inhibitor directly affects catalytic activity. These ionizable groups may affect the enzyme activity by influencing the binding or catalysis or both. Almost all protonation– deprotonation events occurring in an enzyme reaction are catalyzed. The relevant acid–base groups on the enzyme may be identified with the help of techniques such as:

a. Group specific chemical modifications ,
b. pH dependence of kinetic parameters,
c. Structure–activity correlations through physical techniques like X-ray, nuclear magnetic resonance (NMR) spectroscopy, etc.
d. Site-directed mutation analysis .

All the four approaches give useful complementary information; however, pH-dependence of enzyme kinetics gives the best insight as it reports on the reaction being catalyzed while it occurs. The pH dependence of kinetic parameters provides information on kinetic as well as chemical mechanism. The kinetic aspects will be the main focus here, while the other approaches pertaining to pH are dealt with in later sections .

7.1 ENZYME PH OPTIMUM

Determination of activity as a function of pH is the first and simplest experiment one conducts to determine the effect of H+ ions on the enzyme. In all enzymatic studies, maintaining a well-defined pH (H+ concentration) of the system is very crucial. Since a range of pH values (between 0 and 14 in an aqueous environment) are to be used, more than one kind of buffer ion may be required in the experiment. Effects on the enzyme due to switching of buffer species and change in ionic strength, if any, become important and have to be eliminated. This is achieved by the judicious use of suitable buffers . Most enzymes display a bell-shaped pH-activity curve with maximal activity around neutral pH. However there are enzymes with a pH optimum in the acidic (such as pepsin) and alkaline (such as arginase) range as well (Fig. .1).

The decrease of activity on either side of pH optimum may result from (a) instability of the enzyme and/or (b) changes in the kinetic parameters of the enzyme due to pH. Enzyme stability over broad pH ranges is a desirable feature for industrial application. Information about the stability of the enzyme over the pH range studied is also necessary in designing correct kinetic studies.

In order to establish the pH stability profile, the enzyme is preincubated at different pH values. For each pH of preincubation, aliquots are withdrawn as a function of time, and activity remaining is assayed. In this process it must be ensured that the active enzyme remaining is stable after taking the preincubation sample into assay mixture. A plot of percent activity remaining versus pH (after a fixed time of preincubation) gives a fair idea about enzyme's pH stability. Meaningful pH dependence of enzyme activity is then sought within this range of pH defined for stability.

Experimental determination of pH optimum (plot of pH versus activity; Fig. .1) serves two purposes. It is of practical importance in enzyme assay optimization. Secondly, the ascending and descending limbs of such profiles give some idea about the range of pKas and hence possible ionizable groups involved.

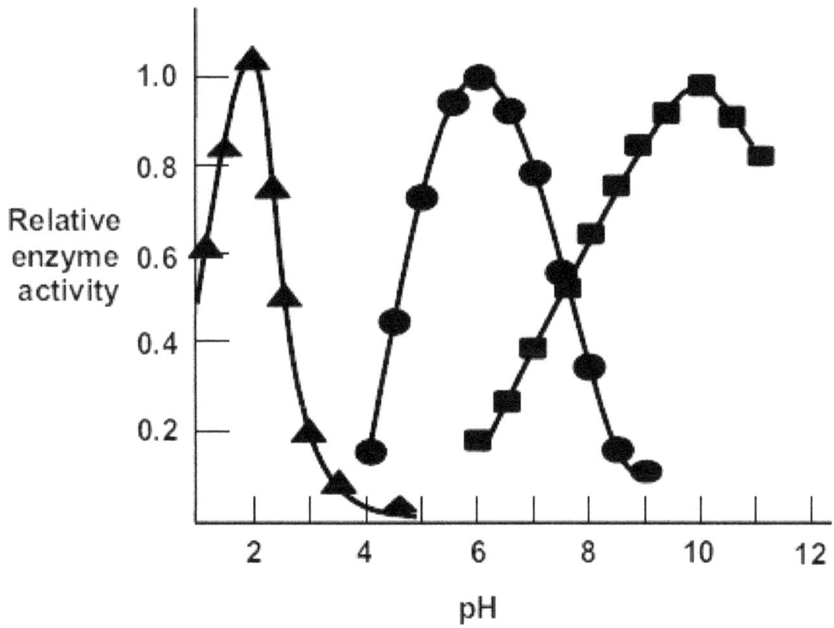

Fig. .1 pH-activity profile of an enzyme. Three representative enzyme examples namely pepsin (~), amylase (●), and arginase (■) are shown

The pH behavior of an enzyme is a complex outcome of all the ionizable groups present on the enzyme and the substrate. A simplistic treatment of bell-shaped activity curve is to compare it to the ionization of a dibasic acid. Examples of enzymes in which at least two active site ionizable groups have been assigned for function are ribonuclease A (H12 and H119) and lysozyme (E35 and D52). Of many ionizable groups present on the enzyme and the substrate, two may be considered kinetically significant with the enzyme represented as HEH. The ascending limb defined by the ionization of the first proton ($HEH \rightleftarrows EH^- + H+$) and the descending limb by the other ($EH^- \rightleftarrows E^{2-} + H+$). The active enzyme species is the singly deprotonated form EH^-. In such a simplistic model representation, the pH behavior of Vmax/KM (the first-order rate constant) is given by the following equation:

$$\frac{V_{\max}}{K_A} = \frac{\left(\frac{V_{\max}}{K_A}\right)^0}{\frac{[H^+]}{K_{H1}} + 1 + \frac{K_{H2}}{[H^+]}}$$

where KH1 and KH2 are the acid dissociation constants for the two groups and (Vmax/ KM)0 is the pH-independent value when the two groups are correctly protonated. At lower pH values (i.e., high [H+]) the KH2/[H+] term in the denominator is insignificant, while at higher pH values (i.e., low [H+]), the [H+]/KH1 term becomes insignificant. Although much oversimplified, this equation does capture the essence of pH dependence of enzyme kinetic parameter (Vmax/KM in this case).

7.2 PH KINETIC PROFILES

Best mechanistic information can be obtained by performing substrate saturation at different pH values. It must be borne in mind, however, that the kinetic mechanism may itself change with change in pH of the reaction. For instance, the normal random mechanism exhibited by creatine kinase changes to an equilibrium ordered one as the pH is decreased from 8.0 to 7.0. Such effects have to be checked up beforehand. The pH-kinetic experiments are so conducted that effects of ionic strength and changing buffer species are properly controlled. With this data one can simultaneously determine the effects of pH on the kinetic constants such as Vmax, Vmax/KM, and also KM. Similarly, one can determine the pH dependence of (a) KI for a competitive inhibitor and (b) Kactivator for an activator. The variation of kinetic parameters withpH is best plotted as log–log plots; this makes sense as pH is a logarithmic scale (-log[H+]). While the horizontal axis is always pH, the Y-axis could be logVmax, logVmax/KM, pKI (-logKI), pKactivator (-logKactivator), pKmetal ion (-logKmetal ion), etc.

Profiles typically schematized to indicate such log–log plots are shown in Fig. .2. Interpretation of pH Profiles Any singly ionizable group can access either a protonated or a deprotonated state. As fractional proton cannot be transferred, one encounters line segments with zero or unit slopes. Rarely, ionization of two protons at a single pH may however appear as lines with a slope of 2. A curve with a unit positive slope followed by a plateau (i.e., zero slope; Fig. .2 top panel) indicates that the protonation of that group results in decreased activity/binding. The point of intersection (extrapolated to pH axis) gives the pKa of the group involved. At higher pH values, a group may lose a proton, and this may lead to decreased activity/ binding (line segment with zero slope followed by one with slope - 1; Fig. .2 middle panel). The presence of both an ascending and a descending limb with a plateau in the middle (Fig. .2 bottom panel) represents an enzyme example with two kinetically significant group protonations. Going from the left to right, deprotonation of the first group increases activity/binding, while the deprotonation of the second group decreases activity/binding. We have already come across an equation describing such a pH behavior of Vmax/KM above. The observed pKa values of the two groups give a glimpse of their possible identity.

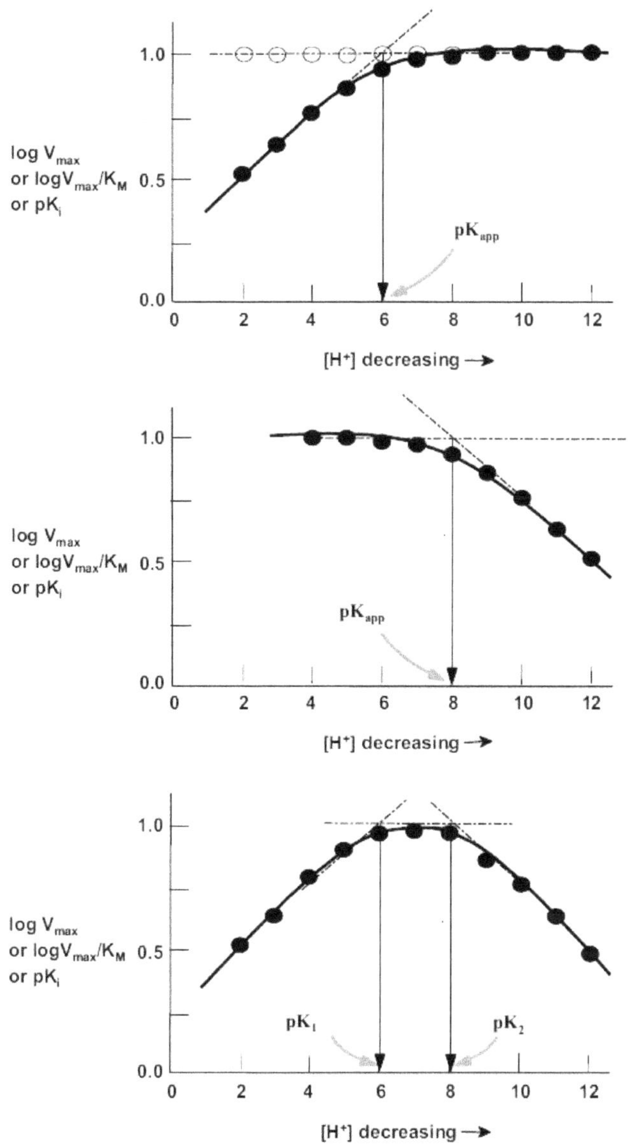

Fig. 2 Possible pH versus kinetic constant profiles encountered for an enzyme. Protonation of the group results in decreased activity/binding (top panel), deprotonation of the group results in decreased activity/binding (middle panel) and combination of the two – indicating titration of two ionizable groups (bottompanel). In each case, the kinetic constant (Vmax, Vmax/KM, or KI) is plotted on the Y-axis as its log value. Open circles (top panel) indicate the profile when that group does not titrate

If one encounters two ionizable groups with their pKa values not sufficiently apart (more than two pH units), then it is difficult to distinguish the two and identify which one is protonated. We can further imagine an unusual situation: going up the pH scale, a group behaving like that in Fig. .2 middle panel may occur before a second group with a profile like that in Fig. .2 top panel. This is an example of a reverse protonation state, and the plateau in Fig. .2 bottom panel is suppressed (to much less than the true value). For an active enzyme, therefore, the group with lower pKa is required in the protonated state, while the group with higher pKa has to be in the deprotonated form. For example, fumarase catalysis requires an active site carboxylate in the deprotonated state and a protonated imidazole. Such reverse protonation cases are not uncommon. It is however conceivable that the catalyst may be locked into an inactive state – possibly because of the proximity of two oppositely charged groups (they may form a salt bridge). Mother nature may have therefore selected suitable pairs of ionizable groups to avoid such thermodynamic wells; lysozyme active site recruits two carboxylates – alternately acting as a general base and general acid for catalysis.

We make a few general observations on the pKas and acid–base groups obtained from pH kinetics studies. Groups that titrate in the logVmax!pH profiles are mainly implicated in the catalytic step, and their ionization occurs in the ES complex (saturating concentration of all reactants). Since Vmax/KM signifies first-order rate constant and interaction of S with E, logVmax/KM ! pH data reflects on essential ionizable groups on the free substrate and the enzyme form with which it combines.

Such groups therefore are important for substrate binding as well as catalysis. This interpretation is further simplified if ionization of acid–base group(s) on the substrate molecule is accounted for or they do not exist (like in glucose). An example of this is glucose isomerase (actually a xylose isomerase from Actinoplanes spp.) and its sitedirected mutant (H54Q) form. This imidazole group (pKa around pH 6.0) of His54 is titrated in the logVmax ! pH profile of the native glucose isomerase but not in the H54Q mutant (the profile with open circles, Fig. .2 top panel).

The Michaelis constant – KM for the substrate – is not an independent kinetic parameter but may be viewed as a ratio of Vmax and Vmax/KM. Therefore its pH dependence is a composite of logVmax ! pH and logVmax/KM ! pH plots. The pKM ! pH profile is rarely plotted. Such profiles are useful indicators of acid–base groups involved in binding (found either on the enzyme or on the substrate) provided KM approximates KS. The simplest pH profile to interpret however is that for the pH dependence of KI for accompetitive inhibitor. In these pKI ! pH profiles, ionizable groups responsible for binding alone are titrated. The KI being equilibrium dissociation constant, the pKa values obtained from such plots are actual pKa values.

They can be meaningfully compared with the group pKas obtained from logVmax!pH and logVmax/KM!pH profiles. pH studies on bovine glutamate dehydrogenase provide an excellent example for such analysis. Glutamate dehydrogenase titrates an ionizable group around pH 8.0 – both in the pKM profile for 2-oxoglutarate (substrate) and pKI profile for oxalylglycine (competitive inhibitor). However, this group is not seen in the pKM profile for 2-oxovalerate – an alternate substrate but missing the γ-COOH (Fig. .3).

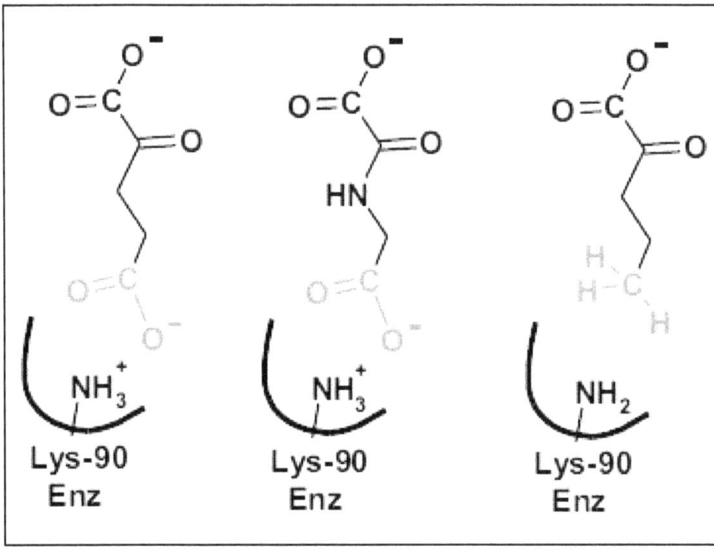

Fig. .3 Interaction of glutamate dehydrogenase with the γ-COOH of 2-oxoglutarate through its active site Lys amino group. Analogous active site interactions with oxalylglycine (competitive inhibitor, middle) and 2-oxovalerate (alternate substrate, right) are also shown

Since –COOH group is fully deprotonated at this alkaline pH, a positively charged group on the enzyme is clearly implicated. Deprotonation of this enzyme Lys side chain adversely affects 2-oxoglutarate binding but not that of 2-oxovalerate; indeed this Lys-NH2 must stay unprotonated for 2-oxovalerate binding. The involvement of this group (K90) anchoring the γ-COOH of 2-oxoglutarate was clearly borne out much later, by X-ray structural data on this enzyme.

Lastly, pKI and pKactivator profiles provide information about the ligands provided by the enzyme that bind the metal ion. The pKa for the ionization of metal ion-bound water may also be seen in such profiles.

7.3 IDENTIFYING GROUPS SEEN IN PH PROFILES

X-Ray, NMR, and chemical modification information on active site functional groups are useful but are limited by the fact that (1) they represent one of the many protein conformations and (2) even when active site pictures with a substrate or inhibitor bound are available, one may be looking at nonproductive complexes. An analog/inhibitor may bind differently from the way a productive substrate interacts at the active site. Site-directed mutations are useful in ruling out important role for a particular group, but mere loss of activity may result from structural changes and not necessarily due to specific role in catalysis.

The pKa values of functionally important acid-base groups are usually significantly shifted from corresponding pKa for groups on free amino acids. Such perturbation of normal pKas is due to unique active site environment - created by the protein - and shielding of these groups from the bulk aqueous phase. Nevertheless, the comparison of experimentally determined enzyme pKa values with reported pKas of amino acid side-chain functional groups can help in identifying their chemical nature. Regardless of our knowledge on exact nature of the acid-base group, its role in catalysis and/or binding can often be assigned from pH-kinetic profiles. Two different techniques may be useful in identifying the functional group whose pKa is observed in enzyme pH profiles.

Inspection of Their Enthalpy of Ionization (ΔHion) The temperature dependence of equilibrium ionization constant is different for different acid-base groups (Table .1). These may be broadly categorized into three groups: groups with low ΔHion (carboxylate and phosphate), groups with moderate ΔHion (imidazole, thiol, and phenolic OH), and groups with high ΔHion (amino, metal-bound water, and guanidinium). In practice, pH profiles are generated at different temperatures, and the pKa values so obtained are plotted against respective 1/T values. The slope of such a plot gives ΔHion/2.303R. Typical ΔHion values of 6 kcal/mol and 12 kcal/ mol correspond to a ΔpKa (between 0 °C and 25 °C) of 0.4 and 0.8 pH units, respectively. A potential identification of groups from their characteristic pKa and ΔHion is thus feasible. A word of caution - unusual ΔHion values - may be obtained when the ionization of the group is coupled with protein conformational change . In such cases experimentally observed ΔHion may include a component for that as well.

Through Solvent Perturbation According to the Bronsted definition of acids and bases, any species of a functional group that has a tendency to lose a proton is an acid. This definition eminently suits our understanding of the role of acid-base groups at the enzyme active site. We can classify all ionizable functionalities on the enzyme (including side chains of amino acid residues) into two groups - cationic acids and neutral acids (Table .2).

Table .1 Range of ΔHion values for ionizable groups on the enzyme

Group	pKa range (at 298 K)	ΔH_{ion} (kcal × mol^{-1})
Carboxylate, phosphate	2–5	0 ± 1.5
Imidazolium, thiol, phenolic OH	5–10	6 ± 1.5
Amine, metal-bound water, guanidinium	8–12	12 ± 1.5

Table .2 Different Bronsted acid groups on the enzyme

Acid type	Functional group
Neutral acids ($X\text{-}H \rightleftarrows X^- + H^+$)	Carboxylate, phosphate, thiol, phenolic OH, metal-bound water
Cationic acids ($X\text{-}H^+ \rightleftarrows X: + H^+$)	Imidazolium, amine, guanidinium

A clear difference between the two Bronsted acid classes is in the net charge generated upon dissociation. Two charges are generated (one proton and a –ve ion) on dissociation of a neutral acid, while there is one +ve charged species on either side of the dissociation equilibrium for a cationic acid (see Table .2 above). This has implications to their differential behavior upon solvent perturbation (altered dielectric constant, etc.). The pKa values of neutral acids are elevated by adding water miscible organic solvents like different alcohols, formamide, dioxane, dimethyl sulfoxide, or N,N-dimethylformamide. Cationic acid pKas are largely unaffected by such solvent addition. In practice, pH profiles are generated with and without the solvent, and the relevant group pKas are compared. The groups whose pKas are elevated in the presence of an added solvent are all neutral acids. The two histidine residues (H12 and H119) at the active site of ribonuclease A were identified by such a study. There are two limitations of solvent perturbation approach. Groups that are not exposed to solvent will not be affected or accounted for. Secondly, since 30–50% of the solvent may have to be used, it becomes important to ensure that the enzyme is not inactivated by such addition.

Chapter 8

ACID–BASE CHEMISTRY AND CATALYSIS

Acids and bases are enormously important in enzyme chemistry. A thorough knowledge of acid–base chemistry is crucial to understand reaction mechanism and catalysis. According to the Bronsted (1923) definition of acids and bases, any substance (or a functional group) that has a tendency to lose a proton is an acid. Correspondingly, a base will then be a proton (H+) acceptor. This definition, as we will see below, eminently suits our understanding of the role of acid–base groups at the enzyme active site. However, Lewis provided a broader definition of acids and bases. Accordingly, a Lewis acid is a substance that accepts an electron pair from a base while a Lewis base is a substance that donates an electron pair to an acid. Lewis acids are involved in enzyme-catalyzed processes as cofactors. Metal cations such as Mg2+, Mn2+, Zn2+, and iron–sulfur clusters are Lewis acids. One way of visualizing this acidity is to consider water molecule coordinated to Zn2+, for instance. Because a lone pair of oxygen is donated to Zn2+ (the Lewis acid), the O-H bond of bound water is better polarized and therefore more readily loses a proton (i.e., H+). While the definition of base is practically same in the two definitions, the concept of Lewis acids is much broader and goes beyond just the H+ donors.

8.1 ACIDS AND BASES

Bronsted Acid–Base and Context of Water The Bronsted–Lowry concept of acids and bases best describes the proton transfers during enzyme catalysis -reactions occurring in water. Every Bronsted acid has a conjugate base associated with it – while every base has a conjugate acid form. Consider the following proton transfer reaction:

$$AH + B \rightarrow A^- + BH^+$$

The acid AH upon losing H+ ion becomes its conjugate base A_. Similarly, the base B by accepting a proton becomes its conjugate acid BH+. Being the universal protic solvent, water can behave as an acid or as a base. It can accept a proton from an acid or donate a proton to a base. Therefore it is no surprise that naked H+ species never occur in aqueous solutions. They invariably are found as solvated hydronium ions (H3O+; and we will always mean H3O+ when we talk of protons in water). The equilibrium extent of these reactions depends on the relative strengths of the acids and bases involved:

$$AH + H_2O \rightleftarrows A^- + H_3O^+$$

$$B + H_2O \rightleftarrows BH^+ + OH^-$$

With a strong acid, water acts as a base and becomes protonated to H3O+. A base on the other hand would deprotonate water to give hydroxide ion (OH_) – here water is acting as an acid. Such compounds that act either as an acid or a base are called amphoteric.

Acids differ in their ability to donate protons. They can be simply compared by measuring the ease with which they transfer a proton to the solvent – water. The strength of an acid in an aqueous solution is expressed by its acid dissociation constant Ka, as defined in the box below. The stronger the acid, the larger is its Ka.

Acid Dissociation Constant

$$Ka = \frac{[H_3O^+] \times [A^-]}{[AH]}$$

This equation can be rearranged to.

$$[H_3O^+] = \frac{Ka \times [AH]}{[A^-]}$$

Taking logarithms on both sides and changing the sign,

$$-\log[H_3O^+] = -\log Ka - \log\frac{[AH]}{[A^-]} = -\log Ka + \log\frac{[A^-]}{[AH]} pH$$
$$= pKa + \log\frac{[A^-]}{[AH]}$$

In order to compare a wide range of acidities (acid strengths), the negative logarithm of Ka (i.e., -log Ka) is more convenient. Obviously, a stronger acid will then have a smaller pKa. It also follows from the Henderson–Hasselbalch equation that when [AH] = [A-], we get pH = pKa. Thus by mapping their pKa values on to the pH scale in water, different Bronsted acids/bases can be compared. It also follows that the ionization of a strong acid generates a weak conjugate base; and the protonated form of a strong base is a weak conjugate acid. Finally, when two bases compete for H+, the stronger base wins the hand of a proton.

Besides being both a proton donor and a proton acceptor, water dissociates toproduce H+ and OH-. The ionization of water due to different states of deprotonation is as shown:

$$H_3O^+ \xrightarrow{pKa=-1.74} H_2O \xrightarrow{pKa=15.74} HO^- \xrightarrow{pKa=21.00} O^{2-}$$

Of these, the first two conjugate acid–base forms of water (namely, H3O+/H2O with a pKa of -1.74 and H2O/OH_ with a pKa of 15.74) are responsible for its amphoteric nature in aqueous solutions. They also define the limits of pH scale in aqueous solutions. The concentration of H+ ions (same as H3O+) in pure water is 10 $^{-7}$ M. In terms of pH scale (negative logarithm of [H+]), this corresponds to a pH of 7.0. In fact, pH=-log ([H+]/1 M) implies that pH is a unitless quantity (for standard state is considered as 1 M). From the concept of pH and the ionic product of water (Kw), the pH scale in aqueous solution spans from 0 to 14. While a pH of 7.0 is neutral (i.e., equal number of H+ and OH- ions), acidic solutions have a pH of less than 7.0; and the lower the pH, the more acidic the solution. Similarly, the higher the pH, the more basic is the solution.

Factors Affecting Bronsted Acid–Base Strength

Broadly three factors influence the strength of a Bronsted acid (or a base).

First, the chemical structure defines the reactivity of a molecule. The number of chlorine atom substitutions on the methyl group of acetic acid influences the acidity. Trichloroacetic acid is a much stronger acid than acetic acid. Similarly, the acidic nature of the phenolic OH increases in the order phenol < p-nitrophenol < picric acid (2,4,6-trinitrophenol). With respect to basicity, aniline is a weaker base than ethylamine, for instance. As a rule, the more stable the conjugate base (e.g., F^- >OH^- > NH_2^-), the stronger the acid (HF > H_2O > NH_3). Also, the more electronegative the element on which the negative charge resides, the more stable is the conjugate base.

Second, the medium by virtue of its solvation effects greatly influences the strength of an acid–base. Dissociation of HCl in gas phase is strongly endothermic – and it does not dissociate. The same molecule readily ionizes in water and is a very strong acid. This is due to favorable solvation of the dissociated charged species.

Consider acetic acid as a solvent in place of water. Since acetic acid is more acidic than water, HCl is a weaker acid when acetic acid is the solvent. Similarly, ammonia is more basic than water, and therefore acetic acid (normally a weak acid) behaves as a strong acid in ammonia. The converse is true for the Bronsted bases. In solvents more acidic than water, all bases behave as stronger bases, and in solvents less acidic than water, they appear as weaker bases. When a base stronger than the hydroxide ion is added to water, it will be converted into its conjugate acid with equimolar release of OH^- ions. A natural outcome of this solvent influence is the leveling effect observed in water. All acids with pKa values below -2.0 (note: pKa of H_3O^+ is -1.74) appear as equally strong acids. This is because, in water, such acids readily protonate the solvent and generate H_3O^+ equivalents. Therefore there cannot be a stronger proton donor than H_3O^+. In the same way, all bases with pKas greater than 16.0 (note: pKa of OH^- is 15.74) appear equally strong because they are virtually completely protonated by the solvent – water. The solvent properties of water are a reflection of its O-H bond strength. A substitution of H by D makes it more rigid and hence most acids are 3–5 times weaker in D_2O than in H_2O.

Lastly, the microenvironment of a Bronsted acid–base greatly influences its ionization and pKa. Consider ionization of the two carboxylate groups of glutamic acid in water. The pKa of the α-COOH is 2.19 while that of its γ-COOH is 4.25. The amino group (α-NH_2) is a good proton acceptor and thus facilitates the ionization of the adjacent α-COOH – its pKa is lowered. The γ-COOH experiences no such effect and hence shows a pKa similar to that of acetic acid (pKa of 4.76). Similarly the α-NH_2 becomes less basic (pKa of 9.67) than the isolated amino group of a primary amine (e.g., ethylamine has a pKa of 10.75). In general, the pKa of a group would be altered if it is involved in a salt bridge formation with an oppositely charged residue or is surrounded by like charges (Fig. .1). Ionization (and separation of charges) is not favored in a hydrophobic environment. The nonaqueous hydrophobic microenvironment (often found at enzyme active sites) destabilizes the charged species and shifts the equilibrium toward a thermodynamically more favorable unionized state.

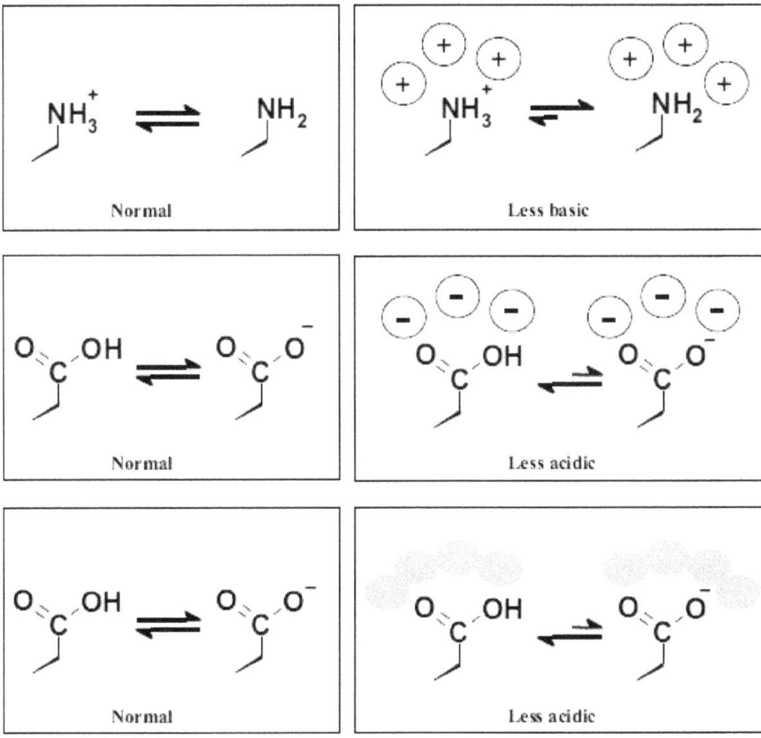

Fig. .1 Perturbation of normal pKa values due to polar and nonpolar microenvironmental effects. Hydrophobic pocket is shown in gray

For instance, acetic acid is a weak acid in benzene than in water; pKa of the -COOH group is elevated in a hydrophobic environment.

All the three factors have been brought to bear by nature in fine-tuning pKa values - in order to optimize acid-base catalysis at enzyme active sites.

Ionizable Groups Relevant to Enzyme Structure and Function

The pH dependence of enzyme activity is a reflection of the acid-base groups involved in substrate binding and catalysis. Ionizable amino acid side chains of the enzyme protein are typically involved in such catalysis . Typical pKa values of such acid- base groups that populate the 0–14 pH range in water are listed in Table .1. Very few acid-base groups other than those of the polypeptide contribute to enzyme function. Water bound to a metal ion (cofactor) on the enzyme can also ionize with characteristic pKa of around 8.0–9.0. The actual value of course will depend on the nature of the metal ion (such as Mg2+, Mn2+, Zn2+or Co2+) and the other ligands it is coordinated to.

The relevance of the acid–base properties of functional groups must be in the context of water and the physiological pH of around 7.0. Enzyme active site bases must therefore be deprotonated around pH 7.0 but have pKa values just below 7.0. The imidazole side chain of histidine residue is one such ideal acid–base group. It is a versatile reagent for enzymatic acid–base chemistry. Many of the pKas listed in Table .1 can be shifted by the microenvironment in the vicinity of the enzyme active site. Because of such perturbations, it is often difficult to unambiguously assign an acid–base group to experimentally observed pKas. For instance, (a) the active site lysine –NH2 of acetoacetate decarboxylase exhibits unusually low pKa of 6.0 , (b) the pKa of the active site glutamate –COOH of lysozyme is attenuated to 6.0 from its typical pKa of 4.3, and (c) papain has a histidine whose imidazole side chain pKa is 3.4. Factors that influence the pKa of such groups include the neighboring charged groups and shielding from bulk solvent by the surrounding hydrophobic residues (e.g., see Fig. .1).

Proton transfer from a group with lower pKa to a group of higher pKa (and not its reverse) is thermodynamically favored. When it is fully transferred to the hydrogen bond acceptor atom, we call it a proton transfer. In a normal hydrogen bond, the H atom is usually found on the hydrogen bond donor but is weakly bonded to the hydrogen bond acceptor atom. Low barrier hydrogen bonds (LBHB) are short, very strong hydrogen bonds – the H atom equally shared between two electronegative atoms of nearly equal pKas. The pKas of the two participating groups have to match for a normal hydrogen bond to become LBHB. Such transitions are demonstrated in many enzymes. This pKa matching, albeit ephemeral, is associated with catalytic events. Thus LBHBs represent one aspect of transient pKa perturbation of acid–base groups that participate in a hydrogen bond.

Some of the groups (included in Table .1) have pKas far from 7.0 and are unlikely to contribute to enzyme catalysis by acting as acid–base groups. These include the guanidinium group of Arg and hydroxyl groups of Ser and Thr. Their participation cannot be categorically ruled out – as their pKas could be sufficiently perturbed through microenvironmental effects. The active site serine-OH of chymotrypsin functions at about 5 pH units below its pKa – around 8.0. However, it acts as nucleophile during catalysis (see next Chapter for details). Guanidinium group of arginine has a very high pKa (>12.0); around neutral pH, where most enzymes function, this group remains fully protonated. It thus cannot be a good acid–base but presents a permanent positive charge. The guanidinium group provides a site through which carboxylate or phosphate can be bound as a cyclic, bi-dentate, hydrogen-bonded ion pair (Fig. .2). Arginine side chain is thus ideally suited to hold substrates in position at the active site. There is growing evidence that guanidinium group of arginine residue may act as a specific anchor for carboxylate (e.g., carboxypeptidase A) and phosphate (e.g., hexokinase) groups in many enzymes.

Table .1 pKa values of Bronsted acid–base groups found on enzymes

Functional group	Structural form: Acid	Structural form: Base	Typical pKa (Range)
Carboxylate			
α-COOH (C-terminal)	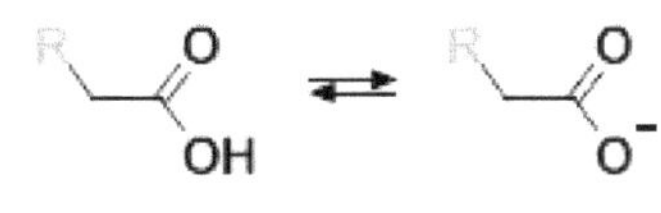		3.4 (2.0–7.0)
β-COOH (Asp)			3.9 (2.0–7.0)
γ-COOH (Glu)			4.3 (2.0–7.0)
Imidazole (His)	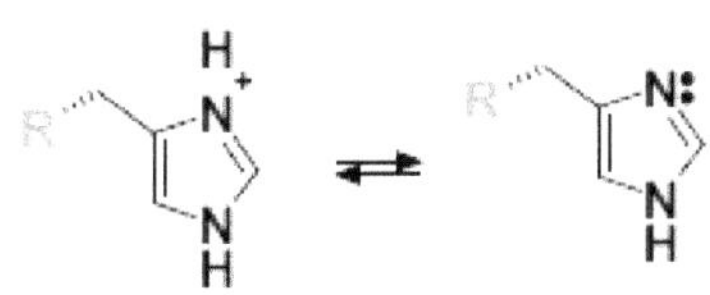		6.5 (6.0–8.0)
α-Amino (N-terminal)	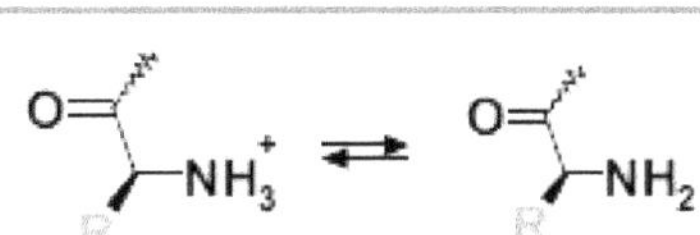		7.5 (6.0–8.0)
ε-Amino (Lys)			10.5 (6.0–10.5)
Sulfhydryl (Cys)			8.3 (7.5–9.0)
Hydroxyl (Tyr)	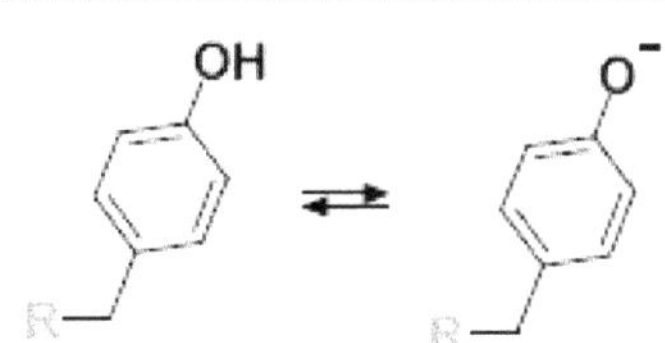		10.0 (8.5–10.5)
Guanidinium (Arg)	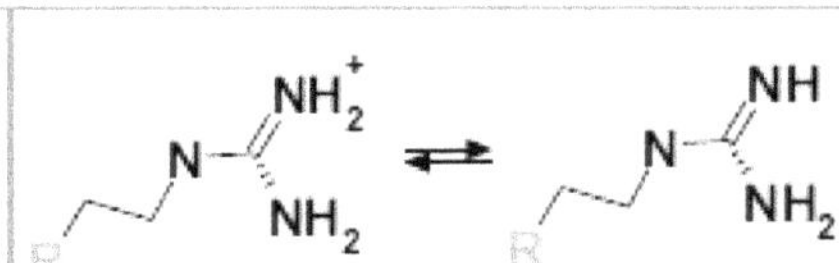		>12.0

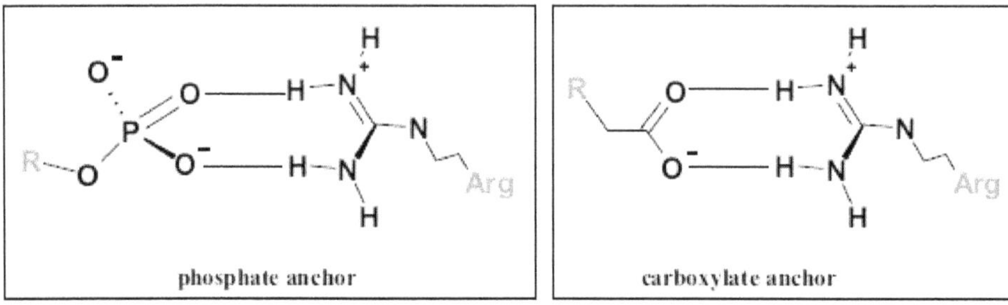

Fig. 2 The guanidinium group of arginine residue provides an excellent site to strongly bind carboxylate and phosphate groups through a cyclic, bifunctional, hydrogen-bonded ion pair (gray lines)

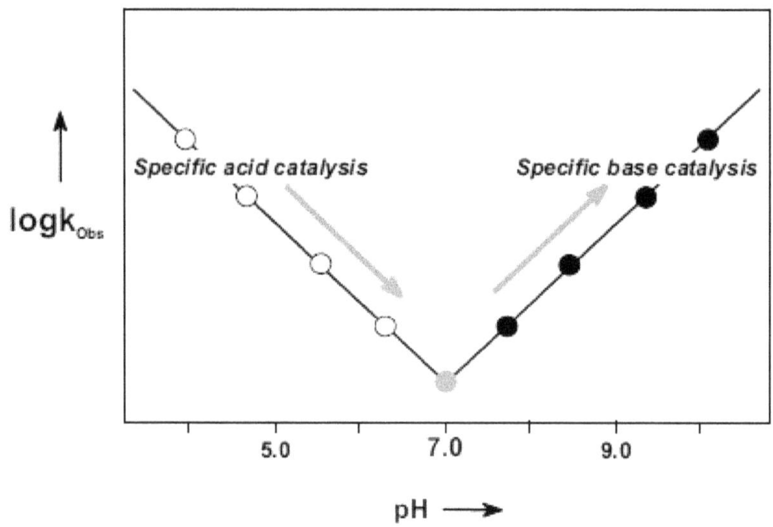

Fig. .3 Influence of pH on the rate of ester hydrolysis in water. The logarithm of kObs is plotted on Y-axis as pH (on the X-axis) is a logarithmic scale

8.2 GENERAL ACID-BASE CATALYSIS

Rates of many reactions are accelerated in the presence of an acid or base. In a simplest form, H+ or OH- may directly participate (act as the acid or the base) in catalysis. When H+ (or H3O+) directly acts as the catalytic acid group, then it is called specific acid catalysis. Similarly, when the catalytic base is OH-, it is termed specific base catalysis. Other components of an aqueous solution (other than H+ or OH-) including buffer species may act as Bronsted acid-base catalyst. Reaction rate acceleration due to a Bronsted acid (other than H+) is referred to as general acid catalysis. Participation of a Bronsted base (other than OH-) similarly defines the general base catalysis.

Hydrolysis of esters is catalyzed by both acids and bases. This reaction is an excellent example of specific acid–base catalysis in water. The contribution of specific acid catalysis decreases with increasing pH while that of specific base catalysis increases with increasing pH. This is expected as [H+] decreases with increasing pH while [OH-] is increasing. Both [H+] and [OH-] are very low at pH = 7.0. Accordingly, the contribution by specific acid–base catalysis is minimal and the experimentally observed hydrolysis rate constant (kObs) is the lowest at pH 7.0 (Fig. .3).

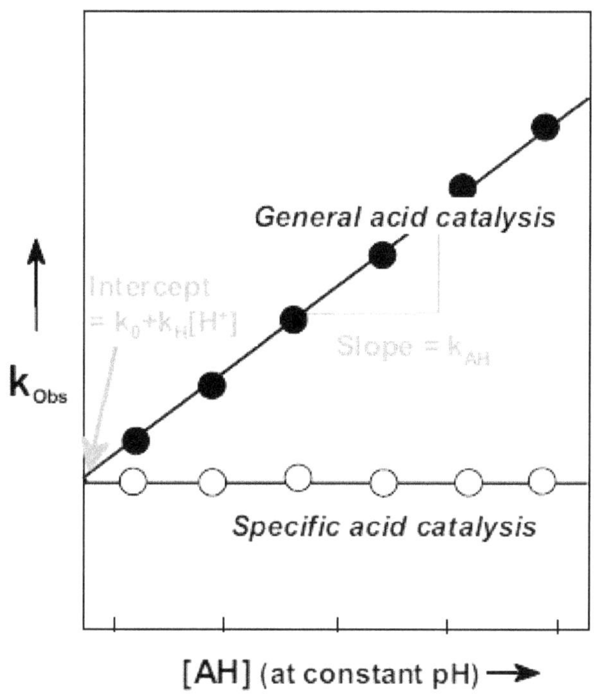

Fig. 4 Effect of buffer (general acid) concentration on the rate of an acidcatalyzed reaction. By simultaneously adjusting [AH] and [A-], the pH is held constant (according to Henderson–Hasselbalch equation) while molarity of the buffer is varied. General acid catalysis will almost always contain the specific acid catalysis component in it (intercept = k0 + kH -[H+])

A general acid (or base) catalysis always comes with the specific acid (or base) catalysis component. Consider a reaction catalyzed by the general acid AH. Since AH dissociates into A- and H+ in water, both AH (a general acid) and H+ (a specific acid) species are available for participation in catalysis. The two contributions can be distinguished by the effects of pH and the acid group concentration (i.e., [AH]) on the rate of chemical reaction. Dependence of the reaction rate on the overall concentration of H+ ions indicates that specific acid catalysis is occurring. We can also conduct the reaction at various buffer strengths but keeping the pH (i.e., [H+]) constant. In such cases, the ionic strength difference can be overcome by suitable salt addition. A linear dependence of reaction rate on the buffer concentration (but the ratio [A-]/[AH] remaining same) indicates general acid catalysis (Fig. .4). Such a rate dependence on buffer concentration is not observed with specific acid catalysis.

Effectiveness of a General Acid Catalyst The effectiveness of a general acid catalyst depends on its acid strength. This is given by the empirical relation known as the Bronsted relationship . According to this relation:

$$\log k_{AH} = \alpha \times \log K_{AH} + C$$

where,

kAH = rate constant of the catalytic step

KAH = dissociation constant (Ka) of the acid AH

α = Bronsted parameter

C = a constant

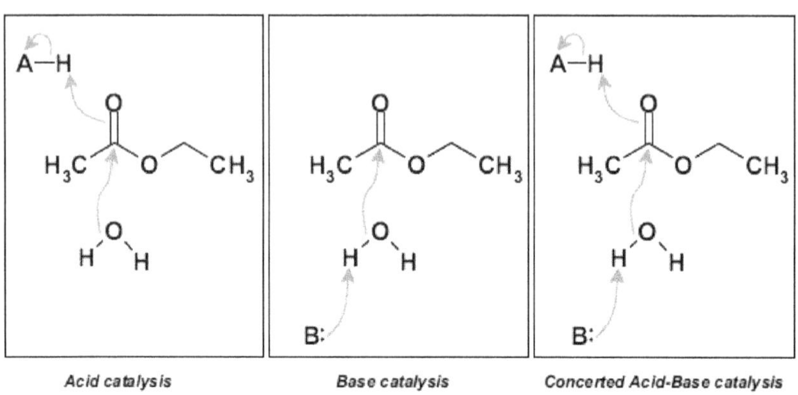

Fig. .5 Acid-base-catalyzed hydrolysis of ethyl acetate. Ester hydrolysis may be facilitated by the acid AH (acid catalysis), the base B: (base catalysis), or both (in concert). In water, when AH = H+, it is specific acid catalysis and when B: = OH-, it is specific base catalysis. Concerted acid-base catalysis involves simultaneous presence and action of a general acid (AH) and a general base (B:). Such an arrangement of groups is best provided at enzyme active sites shielded from the bulk aqueous medium

The Bronsted parameter α indicates the sensitivity of the catalytic step for changes in acid strength (pKa) of AH. A plot of log kAH against log KAH (this is nothing but "-pKa") gives a slope = α. Conceptually, the Bronsted parameter (also known as Bronsted coefficient; normally ranges between 0 and 1) indicates to what extent a proton is transferred from the acid (AH) to the substrate in the transition state. When $\alpha = 1$, every change in acid strength fully affects catalysis. The proton is(almost) completely transferred to the substrate in the transition state (see acid catalysis; Fig. .5). However, when $\alpha = 0$, the reaction is insensitive to changes in acid strength – all acids catalyze the reaction equally strongly (because α - log KAH = 0 and logkAH equals the constant C). Here the proton is hardly transferred in the transition state of the reaction. Lastly, when $\alpha = 0.5$, the proton is transferred halfway between the acid anion A--and the substrate in the transition state –indicating a symmetrical TS for the reaction.

There is also a Bronsted relation for general base catalysis which may be similarly written as logkB: = $-\beta$ XlogK_{BH^+} +C. Here the coefficient β has the same meaning as α for general acid catalysis but a negative sign as we are using the acid dissociation constant for the conjugate acid of the base (B:).

The Hammett equation (log(kx/ko) = ρ X σX) is yet another empirical relationship similar to Bronsted relation. It is more general and may be used to assess the effect of structure (such as any group X) on reactivity; here kx and k0 are rate constants for the reactant structure with substituent X and standard (where X = H), respectively. The Hammett constant ρ (analogous to Bronsted coefficient, α) measures the sensitivity of the reaction to electronic effects. Generally, a positive ρ value means more electrons in the TS than in the reactant and a negative ρ value means fewer electrons in the TS. More details on linear free energy relationships (like Bronsted relationship and Hammett equation) and their utility may be found in specialized texts.

Ester Hydrolysis – An Example Hydrolysis of carboxylate esters is frequently encountered both in organic chemistry and enzymology. Both share acid–base catalysis as a common reaction feature. It is therefore instructive to make a comparison of the two in terms of mechanism of acid–base catalysis. Consider the hydrolysis of ethyl acetate, for instance (Fig. .5). The mechanism of ester hydrolysis involves the formation of a transition state with (a) further polarization of the carbonyl group and (b) partial charge transfer between the ester and water molecule. Such a transition state can be stabilized through proton transfer by an acidic group (AH) to the carbonyl oxygen of the ester. Alternatively, the same transition state may be stabilized by a base (B:) accepting a proton from the attacking water molecule.

Ethyl acetate hydrolysis in water is subject to specific acid (where AH is H+ and $k_{Obs} = k_0 + k_H$-[H+]) or specific base (where B: is OH- and $k_{Obs} = k_0 + k_{OH}$-[OH-]) catalysis. Therefore kObs will not contain contributions due to general acid or general base catalysis (kAH or kB: terms, respectively; see box above). Ester hydrolysis at enzyme active sites is different, however. Since both [H+] and [OH-] are quite small at physiological (near neutral) pH, the contribution of specific acid–base catalysis is marginal. Active sites of esterases (such as lipases, cutinases, and acetyl cholinesterase) contain general acid–base groups for catalysis. Simultaneous and appropriate positioning of both a general acid and a general base greatly facilitates the reaction (Fig. .5). Thus enzyme-catalyzed ester hydrolysis is an excellent example of general acid–base catalysis.

Chapter 9

REVERSIBLE INHIBITORS

The activity of an enzyme can be blocked in a number of ways. For example, inhibitory molecules can bind to sites on the enzyme that interfere with proper turnover. For product inhibition, the product molecule bears some structural resemblance to the substrate and can thus bind to the active site of the enzyme. Product binding blocks the binding of further substrate molecules. This form of inhibition, in which substrate and inhibitor compete for a common enzyme species, is known as *competitive inhibition*. Perhaps less intuitively obvious are processes known as *noncompetitive* and *uncompetitive inhibition*, which define inhibitors that bind to distinct enzyme species and still block turnover. In this chapter, we discuss these varied modes of inhibiting enzymes and examine kinetic methods for distinguishing among them.

There are several motivations for studying enzyme inhibition. At the basic research level, inhibitors can be useful tools for distinguishing among different potential mechanisms of enzyme turnover, particularly in the case of multisubstrate enzymes . By studying the relative binding affinity of competitive inhibitors of varying structure, one can glean information about the active site structure of an enzyme in the absence of a high resolution three-dimensional structure from x-ray crystallography or NMR spectroscopy.

Inhibitors occur throughout nature, and they provide important control mechanisms in biology. Associated with many of the proteolytic enzymes involved in tissue remodeling, for example, are protein-based inhibitors of catalytic action that are found in the same tissue sources as the enzymes themselves. By balancing the relative concentrations of the proteases and their inhibitors, an organism can achieve the correct level of homeostasis. Enzyme inhibitors have a number of commercial applications as well. For example, enzyme inhibitors form the basis of a number of agricultural products, such as insecticides and weed killers of certain types. Inhibitors are extensively used to control parasites and other pest organisms by selectively inhibiting an enzyme of the pest, while sparing the enzymes of the host organism. Many of the drugs that are prescribed by physicians to combat diseases function by inhibiting specific enzymes associated with the disease process (see Table 1.1 for some examples). Thus, enzyme inhibition is a major research focus throughout the pharmaceutical industry.

Inhibitors can act by irreversibly binding to an enzyme and rendering it inactive. This typically occurs through the formation of a covalent bond between some group on the enzyme molecule and the inhibitor. We shall discuss this type of inhibition in Chapter 10. Also, some inhibitors can bind so tightly to the enzyme that they are for all practical purposes permanently bound (i.e., their dissociation rates are very slow). These inhibitors, which form a special class known as *tight binding inhibitors*, are treated separately. In their most commonly encountered form, however, inhibitors are molecules that bind reversibly to enzymes with rapid association and dissociation rates. Molecules that behave in this way, known as classical reversible inhibitors, serve as the focus of our attention in this chapter.

Much of the basic and applied use of reversible inhibitors relies on their ability to bind specifically and with reasonably high affinity to a target enzyme. The relative potency of a reversible inhibitor is measured by its binding capacity for the target enzyme, and this is typically quantified by measuring the dissociation constant for the enzyme—inhibitor complex:

$$[E] + [I] \underset{K_d}{\rightleftharpoons} [EI]$$

$$K_d = \frac{[E][I]}{[EI]}$$

The concept of the dissociation constant as a measure of protein—ligand interactions was introduced in Chapter 4. In the particular case of enzyme— inhibitor interactions, the dissociation constant is often referred to also as the inhibitor constant and is given the special symbol K_i . The K_i value of a reversible enzyme inhibitor can be determined experimentally in a number of ways. Experimental methods for measuring equilibrium binding between proteins and ligands, discussed in Chapter 3, include equilibrium dialysis, and chromatographic and spectroscopic methods. New instrumentation based on surface plasmon resonance technology also allows one to measure binding interactions between ligands and macromolecules in real time. While this method has been mainly applied to determining the binding affinities for antigen—antibody and receptor—ligand interactions, the same technology holds great promise for the study of enzyme—ligand interactions as well. For example, this method has already been used to study the interactions between protein-based protease inhibitors and their enzyme targets .

Although these and many other physicochemical methods have been applied to the determination of K_i values for enzyme inhibitors, the most common and straightforward means of assessing inhibitor binding consists of determining its effect on the catalytic activity of the enzyme. By measuring the diminution of initial velocity with increasing concentration of the inhibitor, one can find the relative concentrations of free enzyme and enzyme—inhibitor complex at any particular inhibitor concentration, and thus calculate the relevant equilibrium constant. For the remainder of this chapter, we shall focus on the determination of K_i values through initial velocity measurements of these types.

9.1 EQUILIBRIUM TREATMENT OF REVERSIBLE INHIBITION

To understand the molecular basis of reversible inhibition, it is useful to reflect upon the equilibria between the enzyme, its substrate, and the inhibitor that can occur in solution. Figure 9.1 provides a generalized scheme for the potential interactions between these molecules. In this scheme, K_i is the equilibrium constant for dissociation of the ES complex to the free enzyme and the free substrate, K_i is the dissociation constant for the EI complex, and k_i is the forward rate constant for product formation from the ES or ESI complexes. The factor β reflects the effect of inhibitor on the affinity of the enzyme for its substrate, and likewise the effect of the substrate on the affinity of the enzyme for the inhibitor. The factor β reflects the modification of the rate of product formation by the enzyme that is caused by the inhibitor. An inhibitor that completely blocks enzyme activity will have β equal to zero. An inhibitor that only partially blocks product formation will be characterized by a value of β between 0 and 1. An enzyme activator, on the other hand, will provide a value of β greater than 1. The question is often asked: Why is the constant α the same for modification of K_s and K_i? The answer is that this constant must be the same for both on thermodynamic grounds. To illustrate, let us consider the following set of coupled reactions:

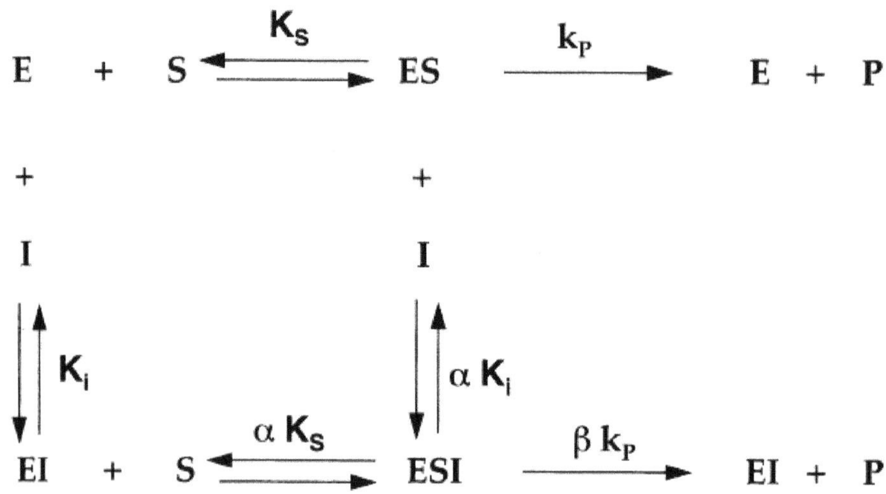

Figure 9.1 Equilibrium scheme for enzyme turnover in the presence and absence of an inhibitor.

$$E + S \underset{K_S}{\rightleftharpoons} ES \qquad \Delta G = RT\ln(K_S) \tag{9.1}$$

$$ES + I \underset{\alpha K_i}{\rightleftharpoons} ESI \qquad \Delta G = RT\ln(\alpha K_i) \tag{9.2}$$

The net reaction of these two is:

$$E + S + I \rightleftharpoons ESI \qquad \Delta G = RT\ln(\alpha K_i K_S) \tag{9.3}$$

Now consider two other coupled reactions:

$$E + I \rightleftharpoons EI \qquad \Delta G = RT\ln(K_i) \tag{9.4}$$

$$EI + S \underset{aK_S}{\rightleftharpoons} ESI \qquad \Delta G = RT\ln(aK_S) \tag{9.5}$$

The net reaction here is:

$$E + S + I \rightleftharpoons ESI \qquad \Delta G = RT\ln(aK_S K_i) \tag{9.6}$$

Both sets of coupled reactions yield the same overall net reaction. Since, as we reviewed ΔG is a path-independent function, it follows that Equations 8.3 and 8.6 have the same value of ΔG. Therefore:

$$RT\ln(\alpha K_i K_S) = RT\ln(aK_S K_i) \tag{9.7}$$

$$\therefore \quad \alpha(K_i K_S) = a(K_i K_S) \tag{9.8}$$

$$\therefore \quad \alpha = a \tag{9.9}$$

Thus, the value of α is indeed the same for the modification of K_s by inhibitorand the modification of K_i by substrate.

The values of α and β provide information on the degree of modification that one ligand (i.e., substrate or inhibitor) has on the binding of the other ligand, and they define different modes of inhibitor interaction with the enzyme.

9.2 MODES OF REVERSIBLE INHIBITION

9.2.1 Competitive Inhibition

Competitive inhibition refers to the case of the inhibitor binding exclusively to the free enzyme and not at all to the ES binary complex. Thus, referring to the scheme in Figure 9.1, complete competitive inhibition is characterized by values of $\alpha=\infty$ and $\beta=0$. In competitive inhibition the two ligands (inhibitor and substrate) compete for the same enzyme form and generally bind in a mutually exclusive fashion; that is, the free enzyme binds either a molecule of inhibitor or a molecule of substrate, but not both simultaneously. Most often competitive inhibitors function by binding at the enzyme active site, hence competing directly with the substrate for a common site on the free enzyme, as depicted in the cartoon of Figure 9.2A. In these cases the inhibitor usually shares some structural commonality with the substrate or transition state of the reaction, thus allowing the inhibitor to make similar favorable interactions with groups in the enzyme active site. This is not, however, the only way that a competitive inhibitor can block substrate binding to the free enzyme. It is also possible (although perhaps less likely) for the inhibitor to bind at a distinct site that is distal to the substrate binding site, and to induce some type of conformation change in the enzyme that modifies the active site so that substrate can no longer bind. The observation of competitive inhibition therefore cannot be viewed as prima facie evidence for commonality of binding sites for the inhibitor and substrate. The best that one can say from kinetic measurements alone is that the two ligands compete for the same form of the enzyme—the free enzyme.

When the concentration of inhibitor is such that less than 100% of the enzyme molecules are bound to inhibitor, one will observe residual activity due to the population of free enzyme. The molecules of free enzyme in this population will turn over at the same rate as in the absence of inhibitor, displaying the same maximal velocity. The competition between the inhibitor and substrate for free enzyme, however, will have the effect of increasing the concentration of substrate required to reach half-maximal velocity. Hence the presence of a competitive inhibitor in the enzyme sample has the kinetic effect of raising the apparent K_m of the enzyme for its substrate without affecting the value of V_{max}; this kinetic behavior is diagnositic of competitive inhibition.

Because of the competition between inhibitor and substrate, a hallmark of competitive inhibition is that it can be overcome at high substrate concentrations; that is, the apparent K_s of the inhibitor increases with increasing substrate concentration.

9.2.2 Noncompetitive Inhibition

"Noncompetitive inhibition" refers to the case in which an inhibitor displays binding affinity for both the free enzyme and the enzyme—substrate binary complex. Hence, complete noncompetitive inhibition is characterized by a finite value of α and β=0. This form of inhibition is the most general case that one can envision from the scheme in Figure 9.1; in fact, competitive and uncompetitive (see below) inhibition can be viewed as special, restricted cases of noncompetitive inhibition in which the value of α is infinity or zero, respectively.

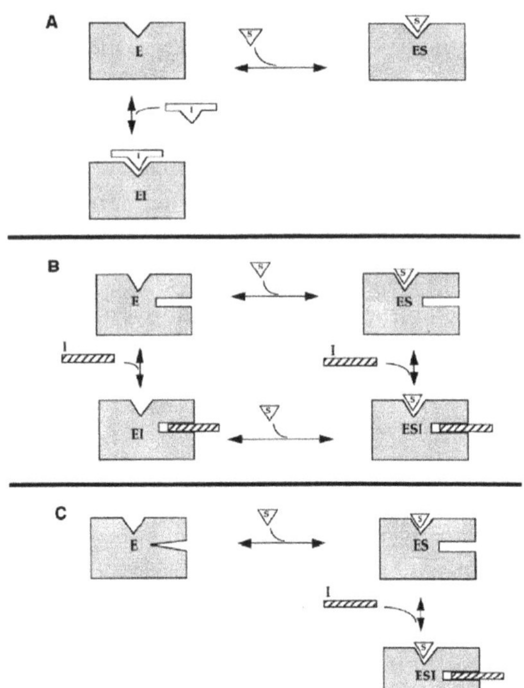

Figure 9.2 Cartoon representations of the three major forms of inhibitor interactions with enzymes: (A) competitive inhibition, (B) noncompetitive inhibition, and (C) uncompetitive inhibition.

Noncompetitive inhibitors do not compete with substrate for binding to the free enzyme; hence they bind to the enzyme at a site distinct from the active site. Because of this, noncompetitive inhibition cannot be overcome by increasing substrate concentration. Thus, the apparent effect of a noncompetitive inhibitor is to decrease the value of V_{max} without affecting the apparent K_m for the substrate. Figure 9.2B illustrates the interactions between a noncompetitive inhibitor and its enzyme target.

The enzymological literature is somewhat ambiguous in its designations of noncompetitive inhibition. Some authors reserve the term "noncompetitive inhibition" exclusively for the situation in which the inhibitor displays equal affinity for both the free enzyme and the ES complex (i.e., $\alpha=1$). When the inhibitor displays finite but unequal affinity for the two enzyme forms, these authors use the term "mixed inhibitors" (i.e., α is finite but not equal to 1).

Indeed, the first edition of this book used this more restrictive terminology. In teaching this material to students, however, I have found that "mixed inhibition" is confusing and often leads to misunderstandings about the nature of the enzyme–inhibitor interactions. Hence, we shall use *noncompetitive inhibition* in the broader context from here out and avoid the term "mixed inhibition." The reader should, however, make note of these differences in terminology to avoid confusion when reading the literature.

9.2.3 Uncompetitive Inhibitors

Uncompetitive inhibitors bind exclusively to the ES complex, rather than to the free enzyme form. The apparent effect of an uncompetitive inhibitor is to decrease V_{max} and to actually decrease K_m (i.e., increase the affinity of the enzyme for its substrate). Therefore, complete uncompetitive inhibitors are characterized by $\alpha<1$ and $\alpha=0$ (Figure 9.2C).

Note that a truly uncompetitive inhibitor would have *no* affinity for the free enzyme; hence the value of K_i would be infinite. The inhibitor would, however, have a measurable affinity for the ES complex, so that αK_i would be finite.Obviously this situation is not well described by the equilibria in Figure 9.1.

For this reason many authors choose to distinguish between the dissociation constants for [E] and [ES] by giving them separate symbols, such as K_{iE} and K_{iES}, K_i and K_1, and K_{is} and K_{11} (the subscripts in this latter nomenclaturerefer to the effects on the slope and intercept values of double reciprocal plots,respectively). Only rarely, however, does the inhibitor have no affinity whatsoeverfor the free enzyme. Rather, for uncompetitive inhibitors it is usually the case that $K_{iE} K_{iES}$. Thus we can still apply the scheme in Figure 9.1 with the condition that $\alpha<1$.

9.2.4 Partial Inhibitors

Until now we have assumed that inhibitor binding to an enzyme molecule completely blocks subsequent product formation by that molecule. Referring to the scheme in Figure 9.1, this is equivalent to saying that $\beta=0$ in these cases. In some situations, however, the enzyme can still turn over with the inhibitor bound, albeit at a far reduced rate compared to the uninhibited enzyme. Such situations, which manifest *partial inhibition*, are characterized by $0<\beta<1$. The distinguishing feature of a partial inhibitor is that the activity of the enzyme cannot be driven to zero even at very high concentrations of the inhibitor. When this is observed, experimental artifacts must be ruled out before concluding that the inhibitor is acting as a partial inhibitor. Often, for example, the failure of an inhibitor to completely block enzyme activity at high concentrations is due to limited solubility of the compound. Suppose that the solubility limit of the inhibitor is 10 μM, and at this concentration only 80% inhibition of the enzymatic velocity is observed.

Addition of compound at concentrations higher that 10 μM would continue to manifest 80% inhibition, as the inhibitor concentration in solution (i.e., that which is soluble) never exceeds the solubility limit of 10 μM. Hence such experimental data must be examined carefully to determine the true reason for an observed partial inhibition. True partial inhibition is relatively rare, however, and we shall not discuss it further. A more complete description of partial inhibitors has been presented elsewhere .

9.3 GRAPHIC DETERMINATION OF INHIBITOR TYPE

9.3.1 Competitive Inhibitors

A number of graphic methods have been described for determining the mode of inhibition of a particular molecule. Of these, the double reciprocal, or Lineweaver–Burk, plot is the most straightforward means of diagnosing inhibitor modality. Recall from Chapter 5 that a double reciprocal plot graphs the value of reciprocal velocity as a function of reciprocal substrate concentration to yield, in most cases, a straight line. As we shall see, overlaying the double-reciprocal lines for an enzyme reaction carried out at several fixed inhibitor concentrations will yield a pattern of lines that is characteristic of a particular inhibitor type. The double-reciprocal plot was introduced in the days prior to the widespread use of computer-based curve-fitting methods, as a means of easily estimating the kinetic values K_m and V_{max} from the linear fits of the data in these plots. As we have described , however, systematic weighting errors are associated with the data manipulations that must be performed in constructing such plots.

To avoid weighting errors and still use these reciprocal plots qualitatively to diagnose inhibitor modality, we make the following recommendation. To diagnose inhibitor type, measure the initial velocity as a function of substrate concentration at several fixed concentrations of the inhibitor of interest. To select fixed inhibitor concentrations for this type of experiment, first measure the effect of a broad range of inhibitor concentrations with [S] fixed at its K_m value (i.e., measure the Langmuir isotherm for inhibition (see Section 9.4) at [S]=K_m). From these results, choose inhibitor concentrations that yield between 30 and 75% inhibition under these conditions. This procedure will ensure that significant inhibitor effects are realized while maintaining sufficient signal from the assay readout to obtain accurate data.

With the fixed inhibitor concentrations chosen, plot the data in terms of velocity as a function of substrate concentration for each inhibitor concentration, and fit these data to the Henri—Michaelis—Menten equation . Determine the values of K_m^{app} (i.e., the *apparent* value of K_m at different inhibitor concentrations) and V_{max}^{app} directly from the nonlinear least-squares best fits of the untransformed data. Finally, plug these values of K_m^{app} and V_{max}^{app} into the reciprocal equation to obtain a linear function, and plot this linear function for each inhibitor concentration on the same doublereciprocal plot. In this way the double-reciprocal plots can be used to determine inhibitor modality from the pattern of lines that result from varying inhibitor concentrations, but without introducing systematic errors that could compromise the interpretations.

Let's walk through an example to illustrate the method, and to determine the expected pattern for a competitive inhibitor. Let us say that we measure the initial velocity of our enzymatic reaction as a function of substrate concentration at 0, 10, and 25 μM concentrations of an inhibitor, and obtain the results shown in Table 9.1.

If we were to plot these data, , we would obtain a graph such as that illustrated in Figure 9.3A. From the fits of the data we would obtain the following apparent values of the kinetic constants:

$$[I] = 0\ \mu M \qquad V_{max} = 100, \qquad K_m = 10.00\ \mu M$$

$$[I] = 10\ \mu M, \qquad V_{max}^{app} = 100, \qquad K_m^{app} = 30.00\ \mu M$$

$$[I] = 25\ \mu M, \qquad V_{max}^{app} = 100, \qquad K_m^{app} = 60.00\ \mu M$$

Table 9.1 Hypothetical velocity as a function of substrate concentration at three fixed concentrations of a competitive inhibitor

[S] (μM)	Velocity (arbitrary units)		
	[I] = 0	[I] = 10 μM	[I] = 25 μM
1	9.09	3.23	1.69
2	16.67	6.25	3.23
4	28.57	11.77	6.25
6	37.50	16.67	9.09
8	44.44	21.05	11.77
10	50.00	25.00	14.29
20	66.67	40.00	25.00
30	75.00	50.00	33.33
40	80.00	57.14	40.00
50	83.33	62.50	45.46

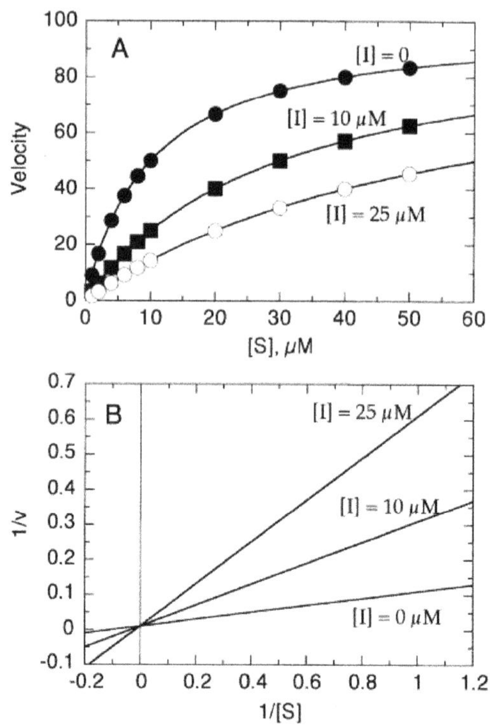

Figure 9.3 Untransformed (A) and double-reciprocal (B) plots for the effects of a competitive inhibitor on the velocity of an enzyme catalyzed reaction. The lines drawn in (B) are obtained by applying Equation 5.24 to the data in (A) and using the apparent values of the kinetic constants in conjunction with Equation 5.34.

If we plug these values of V_{max}^{app} and K_m^{app} into Equation 5.34 and plot the resulting linear functions, we obtain a graph like Figure 9.3B.

The pattern of straight lines with intersecting *y* intercepts seen in Figure 9.3B is the characteristic signature of a competitive inhibitor. The lines intersect at their *y* intercepts because a competitive inhibitor does not affect the apparent value of V_{max}, which, as we saw in Chapter 5, is defined by the *y* intercept in a double-reciprocal plot. The slopes of the lines, which are given by K_m^{app}/V_{max}^{app}, vary among the lines because of the effect imposed on K_m by the inhibitor. The degree of perturbation of K_m will vary with the inhibitor concentration and will depend also on the value of K_i for the particular inhibitor. The influence of these factors on the initial velocity is given by:

$$v = \frac{V_{max}[S]}{[S] + K_m\left(1 + \frac{[I]}{K_i}\right)} \tag{9.10}$$

or, taking the reciprocal of this equation, we obtain:

$$\frac{1}{v} = \frac{1}{V_{max}} + \left(\frac{1}{[S]}\frac{K_m}{V_{max}}\right)\left(1 + \frac{[I]}{K_i}\right) \tag{9.11}$$

Now, comparing Equation 9.11 to Equation 5.34, we see that the slopes of the double-reciprocal lines at inhibitor concentrations of 0 and i differ by the factor $(1+[I]/K_i)$. Thus, the ratio of these slope values is:

$$\frac{\text{slope}_i}{\text{slope}_0} = 1 + \frac{[I]}{K_i} \tag{9.12}$$

or, rearranging:

$$K_i = \frac{[I]}{\left(\frac{\text{slope}_i}{\text{slope}_0}\right) - 1} \tag{9.13}$$

Thus, in principle, one could measure the velocity as a function of substrate concentration in the absence of inhibitor and at a single, fixed values of [I], and use Equation 9.13 to determine the K_i of the inhibitor from the doublereciprocal plots. This method can be potentially misleading, however, because it relies on a single inhibitor concentration for the determination of K_i.

A more common approach to determining the K_i value of a competitive inhibitor is to replot the kinetic data obtained in plots such as Figure 9.3A as the apparent K_m value as a function of inhibitor concentration. The *x* intercept of such a "secondary plot" is equal to the negative value of the K_i, as illustrated in Figure 9.4, using the data from Table 9.1.

In a third method for determining the K_i value of a competitive inhibitor suggested by Dixon , one measures the initial velocity of the reaction as a function of inhibitor concentration at two or more fixed concentrations of substrate. The data are then plotted as $1/v$ as a function of [I] for each substrate concentration, and the value of $-K_i$ is determined from the *x*-axis value at which the lines intersect, as illustrated in Figure 9.5. The Dixon plot ($1/v$ as a function of [I]) is useful in determining the K_i values for other inhibitor types as well, as we shall see later in this chapter.

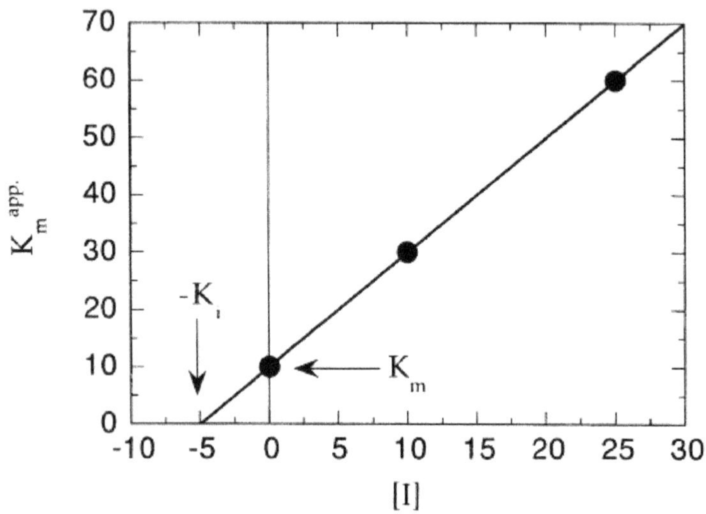

Figure 9.4 Secondary plot of K_m^{app} as a function of inhibitor concentration [I] for a competitive inhibitor. The value of the inhibitor constant K_i can be determined from the negative value of the x intercept of this type of plot.

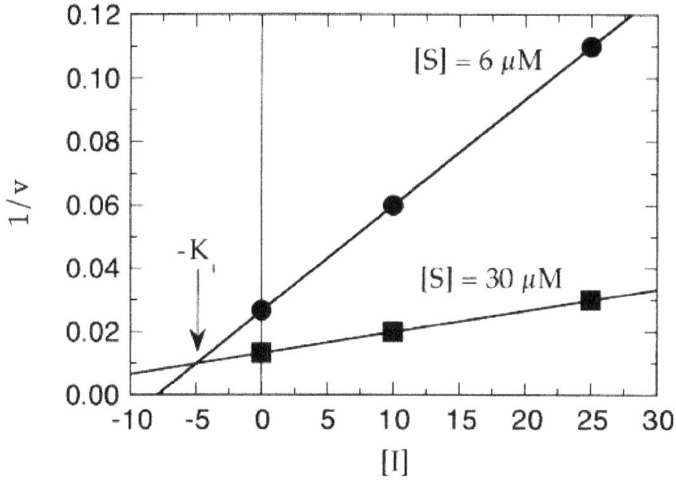

Figure 9.5 Dixon plot (1/v as a function of [I]) for a competitive inhibitor at two different substrate concentrations. The K_i value for this type of inhibitor is determined from the negative of the x-axis value at the point of intersection of the two lines.

9.3.2 Noncompetitive Inhibitors

We have seen that a noncompetitive inhibitor has affinity for both the free enzyme and the ES complex; hence the dissociation constants from each of these enzyme forms must be considered in the kinetic analysis of these inhibitors. The most general velocity equation for an enzymatic reaction in the presence of an inhibitor is:

$$v = \frac{V_{max}[S]}{[S]\left(1 + \frac{[I]}{\alpha K_i}\right) + K_m\left(1 + \frac{[I]}{K_i}\right)} \tag{9.14}$$

and this is the appropriate equation for evaluating noncompetitive inhibitors. Comparing Equations 9.14 and 9.10 reveals that the two are equivalent when α is infinite. Under these conditions the term $[S](1+[I]/\alpha K_i)$ reduces to [S], and Equation 9.14 hence reduces to Equation 9.10. Thus, as stated above, competitive inhibition can be viewed as a special case of the more general case of noncompetitive inhibition.

In the unusual situation that αK_i is exactly equal to K_i (i.e., α is exactly 1), we can replace the term αK_i by K_i and thus reduce Equation 9.14 to the following simpler form:

$$v = \frac{V_{max}[S]}{([S] + K_m)\left(1 + \frac{[I]}{K_i}\right)} \tag{9.15}$$

Equation 9.15 is sometimes quoted in the literature as the appropriate equation for evaluating noncompetitive inhibition. As stated earlier, however, this reflects the more restricted use of the term "noncompetitive."

The reciprocal form of Equation 9.14 (after some canceling of terms) has the form:

$$\frac{1}{v} = \left(1 + \frac{[I]}{K_i}\right)\left(\frac{K_m}{V_{max}}\frac{1}{[S]}\right) + \frac{1 + \frac{[I]}{\alpha K_i}}{V_{max}} \tag{9.16}$$

As described by Equation 9.16, both the slope and the y intercept of the double-reciprocal plot will be affected by the presence of a noncompetitive inhibitor. The pattern of lines seen when the plots for varying inhibitor concentrations are overlaid will depend on the value of α. When α exceeds 1, the lines will intersect at a value of 1/[S] less than zero and a value of 1/v of greater than zero (Figure 9.6A). If, on the other hand, α=1, the lines will intersect below the x and y axes, at negative values of 1/[S] and 1/v (Figure 9.6B). If α=1, the lines converge at 1/[S] less than zero on the x axis (i.e., at 1/[v]=0)

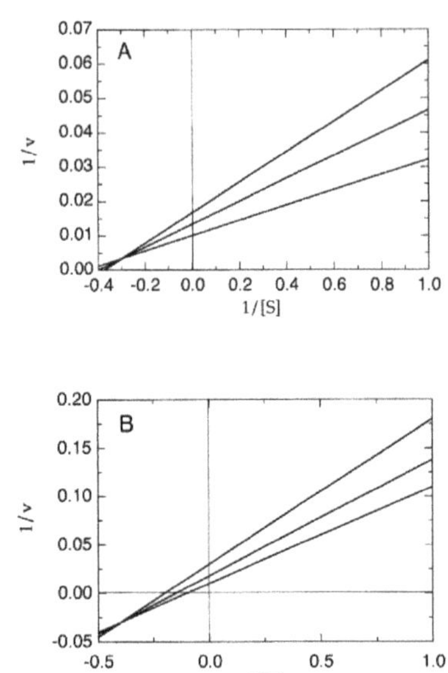

Figure 9.6 Patterns of lines in the double-reciprocal plots for noncompetitive inhibitors for (A)α >1 and (B) α<1.

To obtain the values of K_i and αK_i, two secondary plots must be constructed. The first of these is a Dixon plot of $1/V_{max}$ (i.e., at saturating substrate concentration) as a function of [I], from which the value of αK_i can be determined as the x intercept (Figure 9.7A). In the second plot, the slope of the double-reciprocal lines (from the Lineweaver—Burk plot) are plotted as a function of [I]. For this plot, the x intercept will be equal to αK_i (Figure 9.7B). Combining the information from these two secondary plots allows determination of both inhibitor dissociation constants from a single set of experimental data.

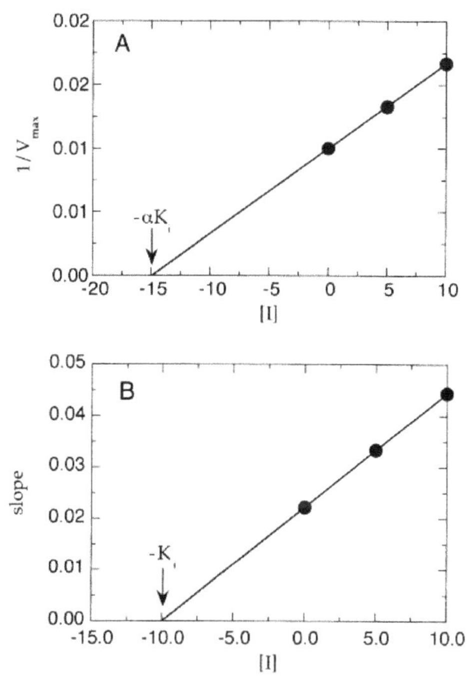

Figure 9.7 Secondary plots for the determination of the inhibitor constants for a noncompetitive inhibitor. (A) $1/V_{max}$ is plotted as a function of [I], and the value of αK_i is determined from the x intercept of the line. (B) The value of αK_i is determined from the x intercept of a plot of the slope of the lines from the double-reciprocal (Lineweaver—Burk) plot as a function of [I].

9.3.3 Uncompetitive Inhibitors

Both V_{max} and K_m are affected by the presence of an uncompetitive inhibitor. The form of the velocity equation therefore contains the dissociation constant αK_i in both the numerator and denominator:

$$v = \frac{\dfrac{V_{max}[S]}{1 + [I]/\alpha K_i}}{\dfrac{K_m}{1 + [I]/\alpha K_i} + [S]} \tag{9.17}$$

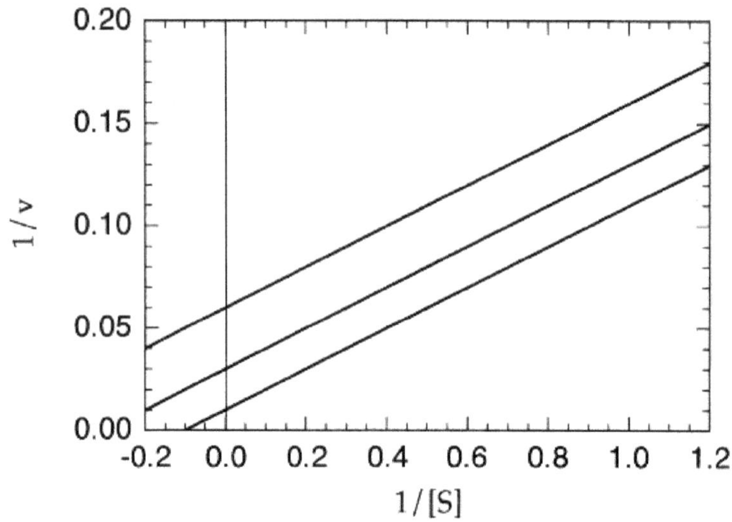

Figure 9.8 Pattern of lines in the double-reciprocal plot of an uncompetitive inhibitor.

If the numerator and denominator of Equation 9.17are multiplied by $(1+[I]/\alpha K_i)$, we can obtain the simpler form:

$$v = \frac{V_{max}[S]}{[S](1 + [I]/\alpha K_i) + K_m} \tag{9.18}$$

The reader will observe that Equation 8.18 is another special case of the more general equation given by Equation 9.14.

With a little algebra, it can be shown that the reciprocal form of Equation 9.17is given by:

$$\frac{1}{v} = \frac{K_m}{V_{max}}\frac{1}{[S]} + \frac{1}{\dfrac{V_{max}}{1 + [I]/\alpha K_i}} \tag{9.19}$$

We see from equation 9.19 that the slope of the double-reciprocal plot is independent of inhibitor concentration and that the *y* intercept increases steadily with increasing inhibitor. Thus, the overlaid double-reciprocal plot for an uncompetitive inhibitor at varying concentrations appears as a series of parallel lines that intersect the *y* axis at different values, as illustrated in Figure 9.8.

For an uncompetitive inhibitor, the *x* intercept of a Dixon plot will be equal to $-\alpha K$ $(1+K_m/[S])$. At first glance this relationship may not look particularly convenient. If, however, one is working at saturating conditions, where $[S]>K_m$, the value of $K_m/[S]$ becomes very small and can be assumed to be zero. Under these conditions, the *x* intercept of the Dixon plot will be equal to αK_i . Thus, *under conditions of saturating substrate*, one can determine the value of αK_i directly from the *x* intercept of a Dixon plot, as described earlier for the case of noncompetitive inhibition.

9.3.4 Global Fitting of Untransformed Data

The best method for determining inhibitor modality and the values of the inhibitor constant(s) is to fit directly and globally all the plots of velocity versus [S] at several fixed inhibitor concentrations to the untransformed equations for competitive (Equation 9.10), noncompetitive (Equation 9.14), and uncompetitive inhibition (Equation 9.18). From analysis of the statistical parameters for goodness of fit (typically χ^2), one can determine which model of inhibitor modality best describes the experimental data as a complete set and simultaneously determine the value of the inhibitor constant(s). This type of global fitting analysis has only recently become widely available. The commercial programs GraphFit and SigmaPlot, for example, allow this type of global fitting [i.e., fitting multiple curves that conform to the functional form $y= f(x, z)$, where *x* is substrate concentration and *z* is inhibitor concentration].

9.4 DOSE--RESPONSE CURVES OF ENZYME INHIBITION

Many biological assays one can measure a specific signal as a function of the concentration of some exogenous substance. A plot of the signal obtained as a function of the concentration of exogenous substance is referred to as a *dose–response plot*, and the function that describes the change in signal with changing concentration of substance is known as a *dose–response curve* (Figure 9.9). These plots have the form of a Langmuir isotherm. We have already seen that such plots can be conveniently used to follow protein–ligand binding equilibria. The same plots are used to follow saturable events in a number of other biological contexts, such as effects of substances on cell growth and proliferation. Dose–response plots also can be used to follow the effects of an inhibitor on the initial velocity of an enzymatic reaction at a fixed concentration of substrate. The concentration of inhibitor required to achieve a half-maximal degree of inhibition is referred to as the IC_{50} value (for inhibitor concentration giving 50% inhibition), and the equation describing the effect of inhibitor concentration on reaction velocity is related to the Langmuir isotherm equation as follows:

$$\frac{v_i}{v_0} = \frac{1}{1 + \frac{[I]}{IC_{50}}} \tag{9.20}$$

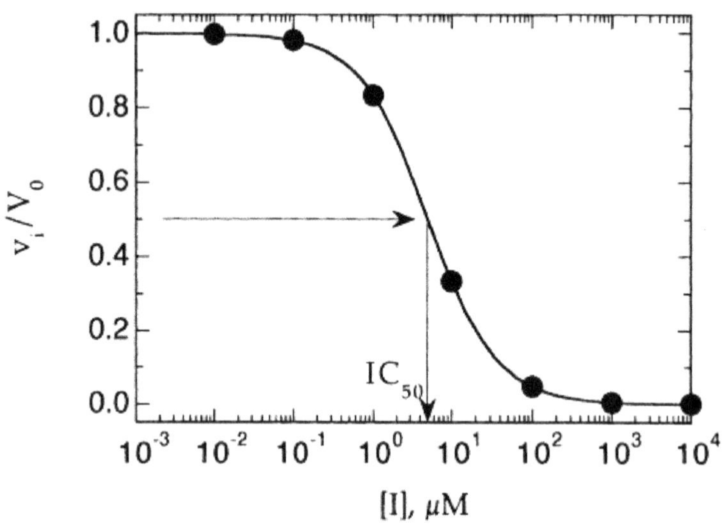

Figure 9.9 Dose–response plot of enzyme fractional activity as a function of inhibitor concentration. Note that the inhibitor concentration is plotted on a log scale. The value of the IC_{50} for the inhibitor can be determined graphically as illustrated.

where v_i is the initial velocity in the presence of inhibitor at concentration [I] and v_0 is the initial velocity in the absence of inhibitor.

The observant reader will note two differences between the form of Equation 9.20 and that of the standard Langmuir isotherm equation . First, we have replaced the dissociation constant K_d (or in the case of enzyme inhibition, K_i) with the phenomenological term IC_{50} . This is because the concentration of inhibitor that displays half-maximal inhibition may be displaced from the true K_i by the influence of substrate concentration, as we shall describe shortly. The second difference between Equations 4.23 and 9.20 is that we have inverted the ratio of [I] and IC_{50} . This is because the standard Langmuir isotherm equation tracks the fraction of ligand-bound receptor molecules. The term v_i/v_0 in Equation 9.20 is referred to as the *fractional activity* remaining at a given inhibitor concentration. This term reflects the fraction of free enzyme, rather than the fraction of inhibitor-bound enzyme.

Considering mass conservation, the fraction of inhibitor-bound enzyme is related to the fractional activity as $1=(v_i/v_0)$. Hence, we could recast Equation 9.20 in the more traditional form of the Langmuir isotherm as follows:

$$\text{fraction bound} = \left(1 - \frac{v_i}{v_0}\right) = \frac{1}{1 + \frac{IC_{50}}{[I]}} \tag{9.21}$$

Dose—response plots are very widely used for comparing the relative inhibitor potencies of multiple compounds for the same enzyme, under well-controlled conditions. The method is popular because it permits analysts to determine the $IC_{50}$by making measurements over a broad range of inhibitor concentrations at a single, fixed substrate concentration. A range of inhibitor concentrations spaning several orders of magnitude can be conveniently studied by means of the twofold serial dilution scheme described in Chapter 5 (Section 5.6.1), with inhibitor being varied in place of substrate here. This strategy is very convenient when many compounds of unknown and varying inhibitory potency are to be screened.

In the pharmaceutical industry, for example, one may wish to screen several thousand compounds as potential inhibitors to find those that have some potency against a particular target enzyme. These compounds are likely to span a wide range of IC_{50} values. Thus, one would set up a standard screening protocol in which the initial velocity of an enzymatic reaction is measured over five or more logs of inhibitor concentrations. In this way the IC_{50} values of many of the compounds could be determined without any prior knowledge of the range of concentrations required to effect potent inhibition of the enzyme.

The IC_{50} value is a practical readout of the relative effects on enzyme activity of different substances under a specific set of solution conditions. In many instances, it is the net effect of the inhibitor on enzyme activity, rather than its true dissociation constant for the enzyme, that is the ultimate criterion by which the effectiveness of a compound is judged. In some situations, a K_i value cannot be rigorously determined because of a lack of knowledge or control over the assay conditions; many times, in these cases, the only measure of relative inhibitor potency is an IC_{50} value. For example, consider the task of determining the relative effectiveness of a series of inhibitors for a target enzyme in a cellular assay. Often, in these cases, the inhibitor is added to the cell medium and the effects of inhibition are measured indirectly by a readout of biological activity that is dependent on the activity of the target enzyme. In a cellular situation like this, one often does not know either the substrate concentration in the cell or the relative amounts of enzyme and substrate (recall that in vitro we set up our steady state conditions so that [S]>[E], but this is not necessarily the case in the cell). Also, in these situations, one does not truly know the effective concentration of inhibitor *within* the cell that is causing the degree of inhibition being measured. This is because the cell membrane may block the transport of the bulk of added inhibitor into the cell. Moreover, cellular metabolism may diminish the effective concentration of inhibitor that reaches the target enzyme.Because of these uncontrollable factors in the cellular environment, often it is necessary to report the effectiveness of an inhibitor as an IC_{50}value.

Despite their convenience and popularity, IC_{50}value measurements can bemisleading if used inappropriately. The IC_{50}value of a particular inhibitor can change with changing solution conditions, so it is very important to report the details of the assay conditions along with the IC_{50}value. For example, in the case of competitive inhibition, the IC_{50}value observed for an inhibitor will depend on the concentration of substrate present in the assay, relative to the K_i of that substrate. This is illustrated in Figure 9.10 for a competitive inhibitor under conditions of [S]=K_m and [S]=10XK_m. Thus, in comparing a series of competitive inhibitors, it is important to ensure that the IC_{50} values are measured at the same substrate concentration. For the same reasons, it is not rigorously correct to compare the relative potencies of inhibitors of different modalities by use of IC_{50} values. The IC_{50} values of a noncompetitive and a competitive inhibitor will vary with substrate concentration, but in different ways. Hence, the relative effectiveness observed in vitro under a particular set of solution conditions may not be the same relative effectiveness observed in vivo, where the conditions are quite different. Whenever possible, therefore, the K_i values should be used to compare the inhibitory potency of different compounds.

It is possible to take advantage of the convenience of IC_{50} measurements and still report inhibitor potency in terms of true K_i values when the mode of inhibition for a series of compounds is known, as well as the values of [S] and K_m. The relationship between the K_i , [S], K_m, and IC_{50} values can be derived from the velocity equations already presented. The derivations have been described in detail by Cheng and Prusoff for competitive, noncompetitive, and uncompetitive inhibitors. The reader is referred to the original paper for the derivations. Here we shall simply present the final forms of the relationships

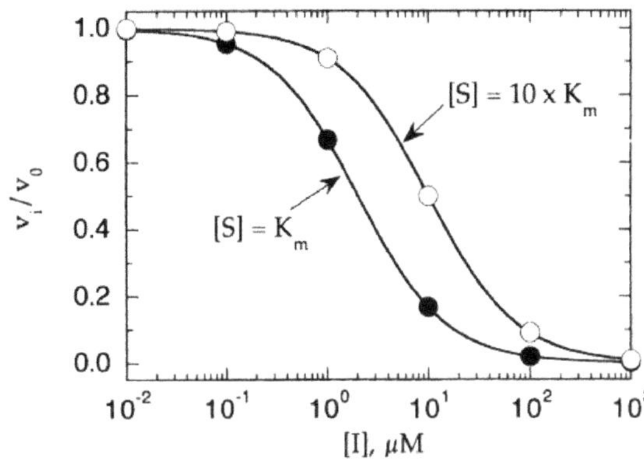

Figure 9.10 Effect of substrate concentration on the IC_{50} value of a competitive inhibitor.

For competitive inhibitors:

$$K_i = \frac{IC_{50}}{1 + \frac{[S]}{K_m}} \tag{9.22}$$

For noncompetitive inhibitors:

$$IC_{50} = \frac{[S] + K_m}{\frac{K_m}{K_i} + \frac{[S]}{\alpha K_i}} \tag{9.23}$$

$$\text{if } \alpha = 1 \quad K_i = IC_{50}$$

For uncompetitive inhibitors:

$$\alpha K_i = \frac{IC_{50}}{1 + \frac{K_m}{[S]}} \qquad (\text{if } [S] \gg K_m, \text{ then } \alpha K_i \sim IC_{50}) \tag{9.24}$$

Equations 9.22—9.24, known as the Cheng and Prusoff relationships, can be conveniently used to convert IC_{50} values to K_i values. To ensure that the correct relationship can be applied, however, it is critical to know the mode of inhibition of the compounds being tested. It might thus seem that there is no great advantage to the use of the Cheng and Prusoff relationships if the mode of inhibition for each compound must be determined by Lineweaver—Burk analysis anyway. In many cases, however, one will wish to measure the relative inhibitory potency of a series of structurally related compounds. If these compounds represent small structural perturbations from a common parent molecule, it is often safe to assume that all the derivative molecules share the same mode of inhibition as the parent. In such situations, one could determine the mode of inhibition for the parent molecule only and then apply the appropriate Cheng and Prusoff relationship to the rest of the molecular series.

There is, of course, the possibility of an inadvertent change in the mode of inhibition as a result of the structural perturbations. This is usually not a great danger if the perturbations are minor, and one can spot-check by performing Lineweaver–Burk analysis on a subgroup of compounds representing a wide range of perturbations within the series. This is a common strategy used in development of structure–activity relationships for the determination of the key structural components in the inhibitory mechanism shared by a series of related compounds, as described next, in Section 9.6. Many scientists, however, consider the K_i values derived by application of the Cheng and Prusoff relationships to be less accurate than those obtained by the more traditional methods described earlier. There is lower confidence in the former results partly because the effects of the inhibitor are examined at only a single, fixed substrate concentration. Nevertheless, because of their convenience, the Cheng and Prusoff relationships are commonly used for high throughput inhibitor screening.

At the beginning of this chapter we mentioned that some inhibitors do not block completely the ability of the enzyme to turnover when bound to the inhibitor. These partial inhibitors will not display the same dose–response curves as full inhibitors because, for these compounds, one can never drive the reaction velocity to zero, even at very high inhibitor concentrations. Rather, the dose–response curve for a partial inhibitor will be best fit by a more generalized form of Equation 9.20, given by:

$$y = \frac{y_{\max} - y_{\min}}{1 + \left(\frac{[\mathrm{I}]}{\mathrm{IC}_{50}}\right)} + y_{\min} \tag{9.25}$$

where y is the fractional activity of the enzyme in the presence of inhibitor at concentration [I], y_{max} is the maximum value of y that is observed at zero inhibitor concentration (for fractional activity, this is 1.0), and y_{min} is the minimum value of y that can be obtained at high inhibitor concentrations.

Unlike the case of full inhibitors, the dose–response curve for a partial inhibitor will reach a minimum, nonzero value of v_i/v_0 at high inhibitor concentrations. In Figure 9.11A, for example, the value of β for our inhibitor is 0.05, so that even at very high inhibitor concentrations, the enzyme still displays 5% of its uninhibited velocity. When behavior of this type is observed, one must be very careful to ensure that the lack of complete inhibition is not an experimental artifact. For example, in densitometry measurements one often observes some finite background density that is difficult to completely subtract out and can give the appearance of partial inhibition when, in fact, full inhibition is taking place.

A more diagnostic signature of partial inhibition can be obtained by arranging the data as a Dixon plot. While all the full inhibitors discussed thus far yielded linear fits in Dixon plots, partial inhibitors typically display hyperbolic fits of the data in these plots (Figure 9.11B). In these cases one can extract the values of α, K_i , and β for the inhibitor, depending on the mode of partial inhibition that is taking place. These analyses are, however, beyond the scope of the present text. The reader who encounters this relatively unusual form of enzyme inhibition is referred to the text by Segel for a more comprehensive discussion of the data analysis.

9.5 MUTUALLY EXCLUSIVE BINDING OF TWO INHIBITORS

If two structurally distinct inhibitors, I and J, are found to both act on the same enzyme, it is possible for them to bind simultaneously to form an EIJ complex (or an ESIJ complex if both inhibitors are capable of binding to the ES complex).

Alternatively, the two inhibitors may bind in a mutually exclusive fashion (i.e., competitive with each other) so that only an EI or an EJ complex can form. There are several tests by which it can be determined if the two inhibitors compete for binding to the enzyme.

The most direct way to measure exclusivity of inhibitor binding is by use of a radiolabeled or fluorescently labeled version of one of the inhibitors. If such labels are used to follow direct binding of the inhibitor to the enzyme, the ability of the second inhibitor to interfere with this binding can be directly measured .

A number of kinetic measures have also been described to test the exclusivity of inhibitor interactions with a target enzyme . All these methods involve measuring the initial velocity of the enzyme at different combinations of the two inhibitors. Theeffects of two inhibitors on the velocity of an enzymatic reaction can be generally described by the following reciprocal relationship:

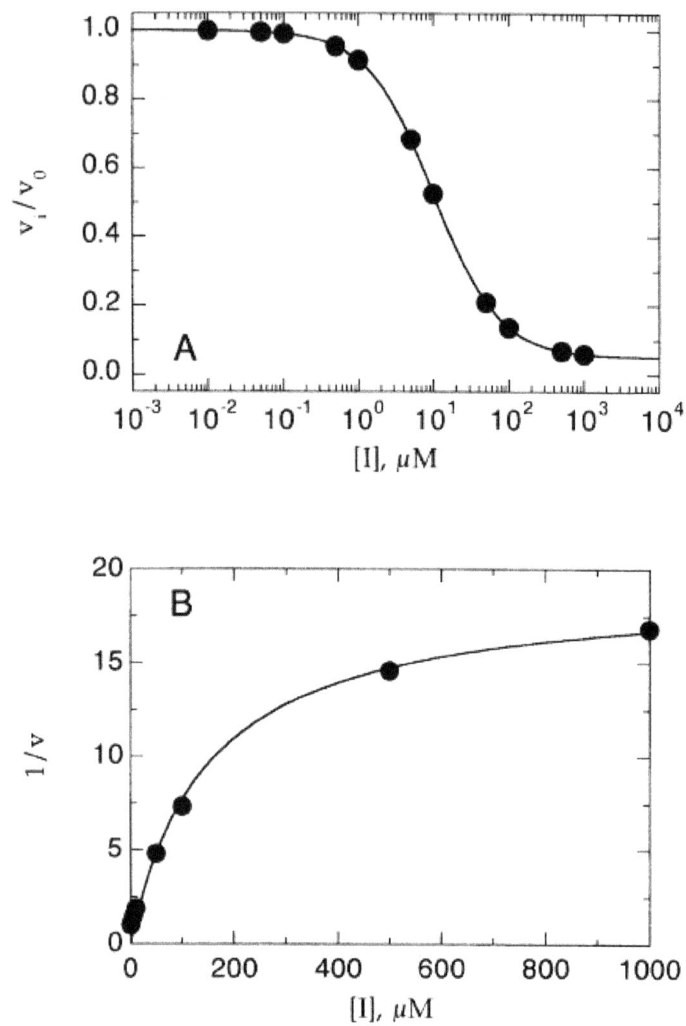

Figure 9.11 Dose–response (A) and Dixon (B) plots for a partial inhibitor. The value of v_i/v_o in (A) reaches a nonzero plateau at high inhibitor concentrations. The hyperbolic nature of the Dixon plot in (B) is characteristic of partial inhibition.

$$\frac{1}{v_{ij}} = \frac{1}{v_0}\left(1 + \frac{[I]}{K_i} + \frac{[J]}{K_j} + \frac{[I][J]}{\alpha K_i K_j}\right) \tag{9.26}$$

where v_{ij} is the initial velocity in the presence of both inhibitors, K_i and K_j are the dissociation constants for inhibitors I and J, respectively, and α is an interaction term that defines the effect of the binding of one inhibitor on the affinity of the second inhibitor. If the two inhibitors bind in a mutually exclusive fashion, $\alpha=\infty$. If the two bind completely independently, $\alpha=1$. If the two inhibitors bind nonexclusively but influence each other's affinity for the enzyme, then α will be finite, but less than or greater than 1. When α is less than 1, the binding of one inhibitor increases the affinity of the enzyme for the second inhibitor, and the binding of the two is said to be *synergistic* (i.e., exhibiting positive cooperativity). When α is greater than 1, the binding of oneinhibitor decreases the affinity of the enzyme for the second inhibitor, and in this case the binding is said to be *antagonistic* (i.e., exhibiting negative cooperativity).

Loewe has described the isobologram method for determining exclusivity of binding. In this analysis different concentrations of I and J are combined to yield the same fractional activity (v_i/v_0). The different concentrations of I in these combinations are plotted on the *y* axis, and the corresponding concentrations of J are plotted on the *x* axis. If the binding of the two inhibitors is mutually exclusive, the data points on such a plot fall on a straight line. If, however, the two inhibitors bind nonexclusively, the data points will form an outwardly concave curve on the isobologram, the curvature depending on the value of α. A number of other graphic methods have been described for this type of analysis ; of all these methods, the most popular is that of Yonetani and Theorell .

In the Yonetani—Theorell method, the data are arranged as Dixon plots, where $1/v_{ij}$ is plotted as a function of [I] at varying fixed concentrations of J. Consideration of Equation 9.26 will reveal that when α is infinity, the data points will form a series of parallel lines when plotted by the method of Yonetani and Theorell (Figure 9.12A). This is an indication that the two inhibitors bind in a mutually exclusive fashion, competing with one another for the same enzyme form. If α is 1, the two inhibitors bind independently, and the lines in the Yonetani—Theorell plot intersect on the *x* axis (Figure 9.12B). If α exceeds 1, the two inhibitors antagonize each other's binding, and the lines on the plot intersect below the *x* axis. Alternatively, if the two inhibitors are synergistic with one another, α is less than 1 and the lines intersect above the *x* axis. For any Yonetani—Theorell plot in which the lines intersect (i.e., $\alpha\neq1$), the *x*-axis value at the point of intersection provides an estimate of $-\alpha K_i$ when [I] is plotted on the *x* axis, or $-\alpha K_i$ when [J] is the variable inhibitor concentration. If the values of K_i and K_j are known from independent measurements, the value of α is then easily calculated.

A common motivation for performing the analysis described in this section is to determine whether two structurally distinct inhibitors share a common binding site on the enzyme molecule. If two inhibitors are found to bind in a mutually exclusive fashion, through either kinetic analysis or direct binding measurements, it is tempting to conclude that they bind to the same site on the enzyme. While this is often true, the caveat described for competitive inhibition with substrate holds here as well: mutually exclusive binding is observed when the two inhibitors bind to a common site on the enzyme, but it can potentially be observed if the two inhibitors bind at independent sites that strongly affect each other through conformational communication, so that ligand binding at one site precludes ligand binding at the second site. Hence, some caution is required in the interpretation of the results of studies of these types.

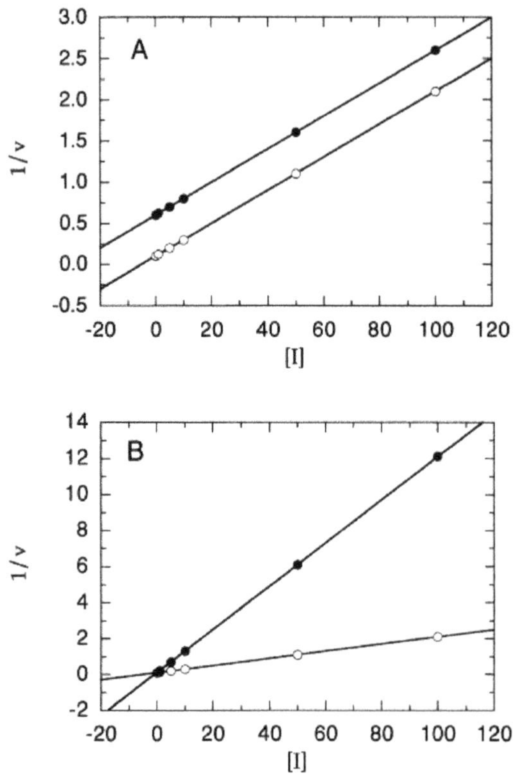

Figure 8.12 (A) Yonetani–Theorell plot for two inhibitors I and J that bind in a mutually exclusive fashion ($\alpha=\infty$) to a common enzyme. (B) Yonetani–Theorell plot for two nonexclusive inhibitors for which $\alpha=1$. Open circles are data points for [J]=0; solid circles are data points for [J]=KJ .

9.6 STRUCTURE–ACTIVITY RELATIONSHIPS AND INHIBITOR DESIGN

Modern attempts to identify inhibitors of specific enzymes have largely focused on elucidating the stereochemical and physicochemical features of inhibitory molecules that allow them to bind well to the enzyme .

Measures of inhibitor potency, such as K_i or IC_{50}, reflect the change in free energy that accompanies transfer of the inhibitor from the solvated aqueous state to the bound state in the enzyme binding pocket (i.e., the ΔG of binding). We have already discussed the types of physicochemical force that are important in protein structure and ligand binding: hydrophobic interactions, hydrogen bonding, electrostatic interactions, and van der Waals forces. The same forces determine the strength of interaction between an inhibitor and an enzyme.

Likewise, we have seen that the shape or topology of a substrate will determine its ability to fit well into the binding pocket of an enzyme, based on the structural complementarity between the enzyme binding pocket and the substrate molecule. It stands to reason that the same structural complementarity should be important in inhibitor binding as well, and this is what is observed empirically. Thus if one can somehow identify a reasonably potent inhibitor of an enzyme, one can begin to make analogues of that molecule with varied structural and physicochemical properties to determine what effect these changes have of inhibitor potency. Attempts to correlate these structural changes with inhibitor potency are referred to as *structure–activity relationship* (SAR) studies.

Today SAR studies can be divided into two major strategic categories: SAR in the absence of structural information on the target enzyme and SAR that utilizes structural information about the enzyme that is obtained from x-ray crystallographic or multidimensional NMR studies. This latter category is also referred to as rational or structure-based inhibitor design.

9.6.1 SAR in the Absence of Enzyme Structural Information

Any SAR study begins with the identification of a lead compound that shows some potency for inhibiting the target enzyme. This lead compound might be identified by random screening of a compound library, such as a natural products library, or it might be based on the known structures of the substrate or product of a particular enzymatic reaction. With a lead compound in hand, analysts subject the substance to small structural perturbations and test these analogues for inhibitor potency. This most basic form of SAR study has been conducted in one way or another since the nineteenth century. The goals of these studies are to determine what structural changes will lead to improved inhibitor potency and to identify the *pharmacophore*, the minimal structure required for inhibition. Once identified, the pharmacophore serves as a template for further inhibitor design.

Consider the enzyme dihydrofolate reductase (DHFR), which catalyzes a key step in the biosynthesis of deoxythymidine. Inhibition of this enzyme blocks DNA replication and thus acts to inhibit cell growth and proliferation. DHFR inhibitors are therefore potentially useful therapeutic agents for the control of aberrant cell growth in cancer, and as antibiotics for the control of bacterial growth. Early attempts to identify inhibitors of this enzyme were based on synthesizing analogues of the substrate dihydrofolate. Figure 9.13 illustrates the chemical structures of dihydrofolate and two classes of DHFR inhibitors; the pteridines, exemplified by methotrexate, and the 5-substituted 2,4-diaminopyrimidines. Methotrexate was identified as a potent inhibitor of DHFR because of its striking structural similarity to the substrate dihydrofolate.

Next, the question of what portions of the methotrexate molecule were critical for DHFR inhibition was addressed by synthesizing various structural analogues of methotrexate. From these studies it was determined that the critical pharmacophore (i.e., the minimal structural component required for inhibition) was the 2,4-diaminopyrimidine ring system. This discovery led to the development of the second class of inhibitors illustrated in Figure 9.13, the 5-substituted 2,4-diaminopyrimidines, of which trimethoprim is a well-known example. Methotrexate is now a prescribed drug for the treatment of human cancers and certain immune-based diseases. Trimethoprim also is a prescribed drug, but its use is in the control of bacterial infections.

An unexpected outcome of the studies on these inhibitors was the finding that the pteridines, such as methotrexate, are potent inhibitors of both mammalian and bacterial DHFR, while trimethoprim and its analogues are much better inhibitors of the bacterial enzymes. The K_i values for trimethoprim for *E. coli*, and human lymphoblast DHFR are 1.35 and 170,000 nM, respectively , a selectivity for the bacterial enzyme of about 126,000-fold! The reason for this spectacular species selectivity was not clear until the crystal structures of mammalian and bacterial DHFR were solved.

Today the enzymologist attempts to develop higher potency inhibitors not simply by random replacement of structural components on a molecule, but rather by systematic and rational changes in stereochemical and physicochemical properties of the substituents. Some properties to be changed are obvious from the structure of the lead compound. If, for example, the lead inhibitor contains a carboxylic acid group, one immediately wonders whether acid— base-type interactions with a group on the enzyme are involved in binding.

One might substitute the carboxylate moiety with an ester, for example, to determine the importance of the carboxylate in binding. More general properties of chemical substituents can be examined as well. These studies call for quantitative measures of the different physicochemical properties to be considered. Among the relevant general properties of chemical substituents, steric bulk, hydrophobicity, and electrophilicity/nucleophilicity are generally agreed

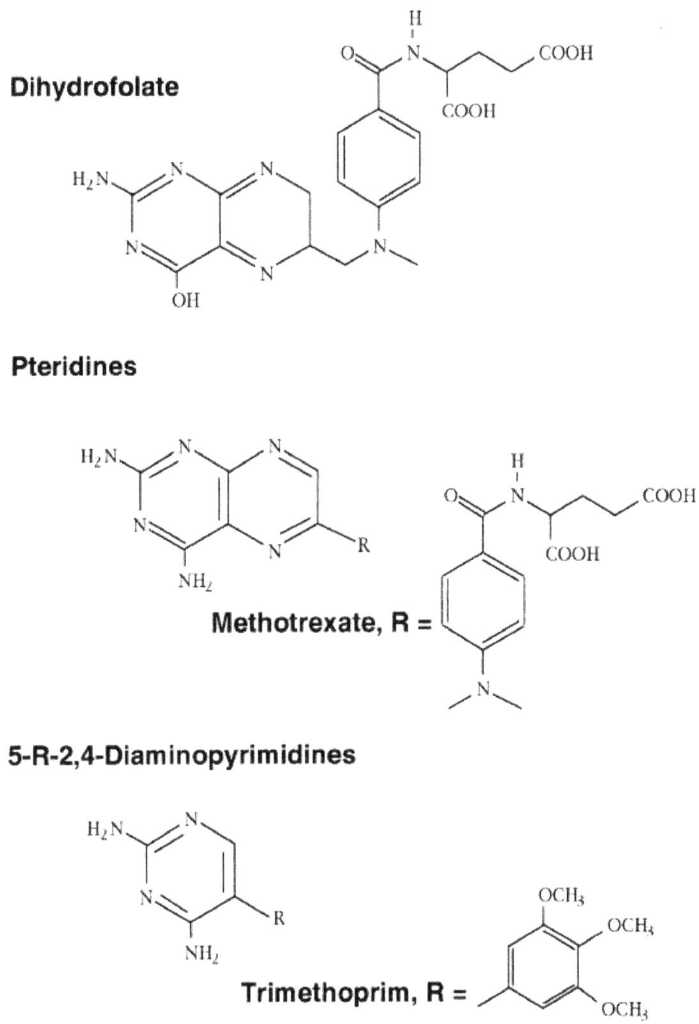

Figure 9.13 Chemical structures of the substrate (dihydrofolate) and two types of inhibitor of the enzyme dihydrofolate reductase (DHFR). to be important factors, and chemists have therefore developed quantitative measures of these parameters.

Several measures have been suggested to quantify steric bulk or molecular volume. One of the earliest attempts at this was the Taft steric parameter E_s , which was defined as the logarithm of the rate of acid-catalyzed hydrolysis of a carboxymethyl-substituted molecule relative to the rate for the methyl acetate analogue :

$$E_s = \log(k_{XCOOCH_3}) - \log(k_{CH_3COOCH_3}) \tag{9.27}$$

A more geometric measure of steric bulk is provided by the Verloop steric parameter, which basically measures the bond angles and bond lengths of the substituent group . Chemists also have used the molar refractivity as a measure of molecular volume of substituents . The molar refractivity, MR, is defined as follows:

$$\mathrm{MR} = \frac{n^2 - 1}{n^2 + 1} \frac{\mathrm{MW}}{d} \tag{9.28}$$

where n is the index of refraction, MW is the molecular weight, and d is the density of the substituent under consideration. Since n does not vary widely among organic molecules, MR is mainly a measure of molecular volume.

The relative hydrophobicity of molecular substituents is most commonly measured by their partition coefficient between a polar and nonpolar solvent. For this purpose, chemists have made water and octanol the solvents of choice. The molecule is dissolved in a 1:1 mixture of the two solvents, and its concentration in each solvent is measured at equilibrium. The partition coefficient is then calculated as the equilibrium constant:

$$P = \frac{[\mathrm{I}]_{\mathrm{octanol}}}{[\mathrm{I}]_{\mathrm{water}}} \tag{9.29}$$

In measuring relative hydrophobicity, the effect of different substituents, on the partition coefficient of benzene in octanol/water is used as a standard. The hydrophobic parameter π is used for this purpose, and is defined as follows :

$$\pi = \log(P_x) - \log(P_H) \tag{9.30}$$

where P_X is the partition coefficient for a monosubstituted benzene with substituent x, and P_H is the partition coefficient of benzene itself.

The most widely used index of electronic effects in inhibitor design is the Hammett σ constant. Originally developed to correlate quantitatively the relationship between the electron-donating or -accepting nature of a parasubstituent on the ionization constant for benzoic acid in water ; this index is defined as follows:

$$\sigma = \log(K_x) - \log(K_H) \tag{9.31}$$

where K_X is the ionization constant for the parasubstituted benzoic acid with substituent x and K_H is the ionization constant for benzoic acid. Groups that are electron acceptors (e.g., COOH, NO_2, NR_3^+) withdraw electron density from the ring system, hence stabilize the ionized form of the acid; such groups have positive values of. Electron-donating groups (e.g., OH, OCH_2, NH_3)have the opposite effect on the ionization constant and thus have negative values of. Values of for a very large number of organic substituents have been tabulated by several authors. One of the most comprehensive list of values can be found in the text by Martin .

The ability to quantify these various physicochemical properties has led to attempts to express the inhibitor potency of molecules as a mathematical function of these parameters. This strategy of *quantitative structure—activity relationships* (QSAR) was first championed by Hansch and his coworkers . In a typical QSAR study, a series of analogues of a lead inhibitor is prepared with substituents that systematically vary the parameters described earlier. The experimentally determined potencies of these compounds are then fit to varying linear and nonlinear weighted sums of the parameter indices to obtain the best correlation by regression analysis. Equations 9.32—9.34 illustrate forms typically used in QSAR work.

$$\log\left(\frac{1}{K_i}\right) = a\pi + b\sigma + \mathrm{cMR} + d \tag{9.32}$$

$$\log\left(\frac{1}{K_i}\right) = a\pi^2 + b\sigma + \mathrm{cMR} + d \tag{9.33}$$

$$\log\left(\frac{1}{K_i}\right) = a\pi^2 + b\sigma + \mathrm{c}\log(\beta\mathrm{MR}) + d \tag{9.34}$$

Equation 9.32 is a simple linear relationship, while Equations 9.33 and 9.34 have nonlinear components. In these equations the values *a*, *b*, *c*, *d*, and β are proportionality constants, determined from the regression analysis. In developing a mathematical expression for the correlation relationship here, one hopes to predict the inhibitory potency of further compounds prior to their synthesis, based on the equation established from the QSAR. In practice, the predictive power of these QSAR equations varies dramatically. When such predictions fail, it is usually because additional factors that influence inhibitor potency were not quantitatively included in the functional expression. In some cases these additional factors are neither well understood nor easily quantified.

As a simple example of QSAR, let us again consider the inhibition of bacterial DHFR by pteridines and 5-substituted 2,4-diaminopyrimidines. Coats et al. studied the QSARs of both classes of compounds for their ability to inhibit the DHFR from the bacterium *L actobacillus casei*. They measured the IC values for 25 pteridine analogues with different *R* substituents and also for 33 5-substituted 2,4-diaminopyrimidine analogues with different *R* groups (Figure 9.13). From these data they determined the QSAR equations given by Equations 9.35 and 9.36 for the pteridines and 5-*R*-2,4-diaminopyrimidines, respectively:

$$\log\left(\frac{1}{IC_{50}}\right) = 0.23\pi - 0.004\pi^2 + 0.77I + 3.39 \tag{9.35}$$

$$\log\left(\frac{1}{IC_{50}}\right) = 0.38\pi - 0.007\pi^2 + 0.66I + 2.15 \tag{9.36}$$

In these equations, the π parameter refers to the hydrophobicity of the *R* group, and the index *I* is an empirical parameter related to the presence of a —N—C— or —C—N— bridge between the parent ring system and an aromatic ring on the substituent . The relationships between the ICvalues calculated from these equations and the experimentally determined ICvalues are illustrated in Figure 9.14. Again, one must keep in mind that the correlations illustrated are for the molecules used to establish the QSAR equations. The value of these equations in predicting the inhibitor potency of other molecules will depend on how significantly other unaccounted- for factors influence potency. Nevertheless, QSAR provides a means of rationalizing the observed potencies of structurally related compounds in terms of familiar physicochemical properties. An up-to-date volume by Kubinyi provides a detailed and practical introduction to the field of QSAR. This text should be consulted as a starting point for acquiring a more in-depth treatment of the subject.

Another approach to designing potent inhibitors of enzymes is to considernthe probable structure of the transition state of the chemical reaction catalyzed by the enzyme. As described , the catalytic efficiency of enzymes is due largely to their ability to achieve transition state stabilization. If this stabilization is equated with binding energy, a stable analogue that mimics the structure of the transition state should bind to an enzyme some 10^{10}—10^{15} times greater than the corresponding ground state substrate molecule . Since typical substrate K_m values are in the millimolar-tonanomolar range, a true transition state mimic may bind to its target enzyme with a K_i value between 10^{-13} and 10^{-24} M! With such incredibly tight binding affinities, such inhibitors would behave practically as irreversible enzyme inactivators.

The foregoing approach to inhibitor design has been hindered by the great difficulty of obtaining information on transition state structure by traditional physical methods. Because they are so short-lived, the transition state species of most enzymatic reactions are present under steady state conditions at very low concentrations (i.e., femtomolar or less). Hence, attempts to obtain structural information on these species from spectroscopic or crystallographic methods have been largely unsuccessful. Information on transition state structure can, however, be gleaned from analysis of kinetic isotope effects on enzyme catalysis, as recently reviewed by Schramm et al. . As discussed , kinetic isotope effects are observed because of the changes in vibrational frequencies for the reactant and transition state species that accompany heavy isotope incorporation. By synthesizing substrate analogues with heavy isotopes at specific locations, one can determine the kinetic isotope effects imparted by each replacement. From this type of information, one can use vibrational normal mode calculations to identify the vibrational modes that are most strongly perturbed in the transformation from reactant to transition state of the substrate, hence to map out the structural changes that have occurred in the molecule. The information thus obtained can then be used to design molecules that mimic the structure of the reaction transition state.This approach has been applied to the design of transition state analogues of renin, an aspartyl protease .

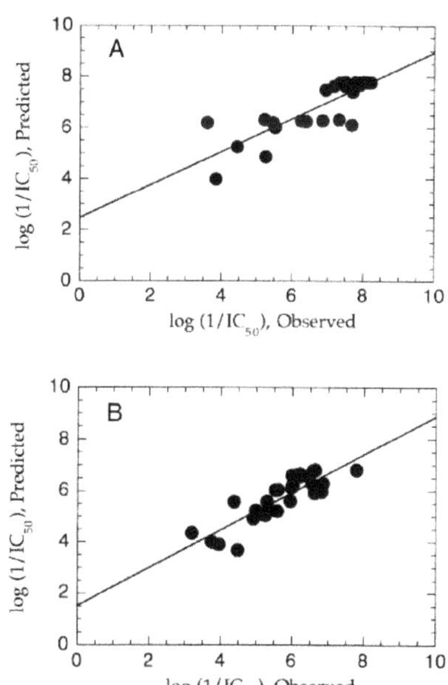

Figure 9.14 QSAR correlation plots for the potencies of pteridines (A) and 5-substituted 2,4-diaminopyrimidines (B) as inhibitors of the dihydrofolate reductase from L. casei.

In vivo, renin is responsible for the proteolytic processing of angiotensinogen to angiotensin I, the progenitor of the vasoconstrictor peptide angiotensin II. The substrate is hydrolyzed at a Leu-Val peptide bond, and the hydrolysis reaction is proposed to utilize an active site water molecule as the attacking nucleophile to produce the tetrahedral transition state illustrated in Figure 9.15A. The peptide sequence of the renin substrate angiotensinogen is shown in Figure 9.15C. In their first attempt at an inhibitor of renin, Blundell and coworkers replaced the P1 carbonyl group by a methylene linkage, yielding the reduced isostere (—CH_2—NH—) containing peptide inhibitor 1 (Figure 9.15C). The investigators next noted that the pepstatins are naturally occurring protease inhibitors that contain the unusual amino acid statine (Figure 9.15B), which in turn contains a —CH(OH)— moiety that resembles the proposed transition state of renin. Pepstatin is a poor renin inhibitor; but, reasoning that the statine group was a better transition state mimic than the reduced peptide isostere (—CH_2 —NH—), Blundell et al. incorporated this structure into their peptide inhibitor to produce 2 (Figure 9.15C). The closest analogue to the true transition state of the reaction would be one incorporating

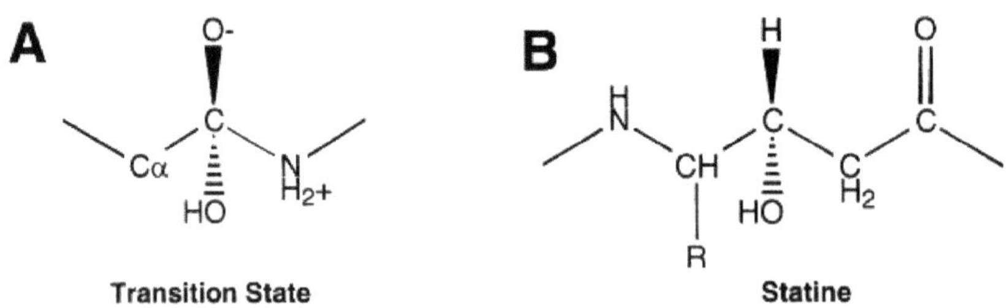

C

Compound	P1		P1'	IC_{50} (μM)
Substrate	Ile His Pro Phe His Leu-	CO - NH-	Val Ile His Asn	-
1	Pro His Pro Phe His Leu-	CH_2 -NH-	Val Ile His Lys	0.01
2	Boc His Pro Phe His Leu- CH(OH)-CH_2-CO-NH-Leu Phe			0.01
3	Boc His Pro Phe His Leu-	CH(OH)-CH_2-	Val Ile His	0.0007

Figure 9.15 (A) Proposed structure of the tetrahedral transition state of the renin proteolysisnreaction. (B) Chemical structure of the statine moiety of natural protease inhibitors, such as npepstatin. (C) Structures and IC_{50} values for the peptide substrate and inhibitors of renin, incorporating various forms of transition state analogues.

a —CH(OH)—NH— group at P1—P1' of the peptide. Since, however, synthesis of this transition state analogue was hampered by the instability of the resulting compound, Blundell's group instead synthesized a closely related analogue containing a hydroxyethylene moiety [(—CH(OH)—CH —), 3, which proved to be an extremely potent inhibitor of the enzyme (Figure 9.15C).

To date, the de novo design of transition state analogues as enzyme inhibitors has been applied only to a limited number of enzymes by a handful of laboratories. With improvements in the computational methods associated with this strategy, however, more widespread use of this approach is likely to be seen in the future.

9.6.2 Inhibitor Design Based on Enzyme Structure.

In the search for potent enzyme inhibitors, knowledge of the three dimensional structure of the inhibitor binding site on the enzyme provides the ultimate guide to designing new compounds. The structures of enzyme active sites can be obtained in atomic detail from x-ray crystallography and multidimensional NMR spectroscopy. A detailed discussion of these methods is beyond the scope of the present text. Our discussion will focus instead on the use of the structural details obtained from these techniques. The reader interested in learning about protein crystallography and NMR spectroscopy can find many excellent review articles and texts .

The crystal or NMR structure of an enzyme with an inhibitor bound provides structural details at the atomic level on the interactions between the inhibitor and the enzyme that promote binding. Hydrogen bonding, salt bridge formation, other electrostatic interactions, and hydrophobic interactions can be readily inferred from inspection of a high resolution structure. Figure 9.16 provides simplified schematic representations of the binding interactions between DHFR and its substrate dihydrofolate and inhibitor methotrexate, illustrating the involvement of common amino acid residues in the binding of both ligands. These structural diagrams also indicate that the orientation and hydrogen bonding patterns are not identical for the substrate and the inhibitor.

Nevertheless, the major forces involved in binding of both ligands to the enzyme are hydrogen bonds between amino acid residues of the active site and the 2,4-diaminopyrimidine ring of the ligands. Visual inspection by means of molecular graphics methods suggested that this ring constituted the critical pharmacophore and led to the design of trimethoprim, the prototypical 5-substituted 2,4-diaminopyrimidine . As we have seen, these structural inferences are consistent with the SAR and QSAR studies of DHFR inhibitors.

The example of trimethoprim suggests a straightforward, if tedious, means of utilizing structural information in the design of new enzyme inhibitors: namely, the iterative design, synthesis, and crystallization of inhibitor—enzyme complexes. In this approach, one starts with the crystal structure of the free enzyme or of the enzyme—lead inhibitor complex. Based on inspection of the crystal structure, one suggests, changes in chemical structure of the inhibitor to better engage the enzyme active site. The new compound is then synthesized and tested for inhibitory potency. Next, to determine whether the predicted interactions in fact occur, a crystal structure of the enzyme with this new inhibitor bound is obtained. This new structure is then used to search for additional changes to the inhibitor structure that might further improve potency, and the process is continued until an inhibitor of sufficient potency is obtained. This iterative structure-based inhibitor design method was used in the design and synthesis of inhibitors of thymidylate synthase reported by Appelt et al. ; this paper provides a good illustration of the method.

Thus, the first step to structure-based inhibitor design is to obtain a crystal or NMR structure of the target enzyme, with or without a lead inhibitor bound to it. In some cases, the determination of a crystal or NMR structure of the target protein proves problematic because of the technical difficulties associated with crystallographic and NMR methods. If the structure of a closely related enzyme has been reported, however, one can still attempt to model the three-dimensional structure of the target enzyme by means of homology . In homology modeling one attempts to build a model of the target enzyme by superimposing the amino acid residues of the target onto the three-dimensional structure of the homologous protein whose structure has been solved. For these method to work, the target enzyme and its homologue must share at least 30% amino acid sequence identity. The accuracy of the model obtained in this way is directly related to the degree of sequence identity between the two proteins: the greater the sequence identity, the more accurate the modeled structure.

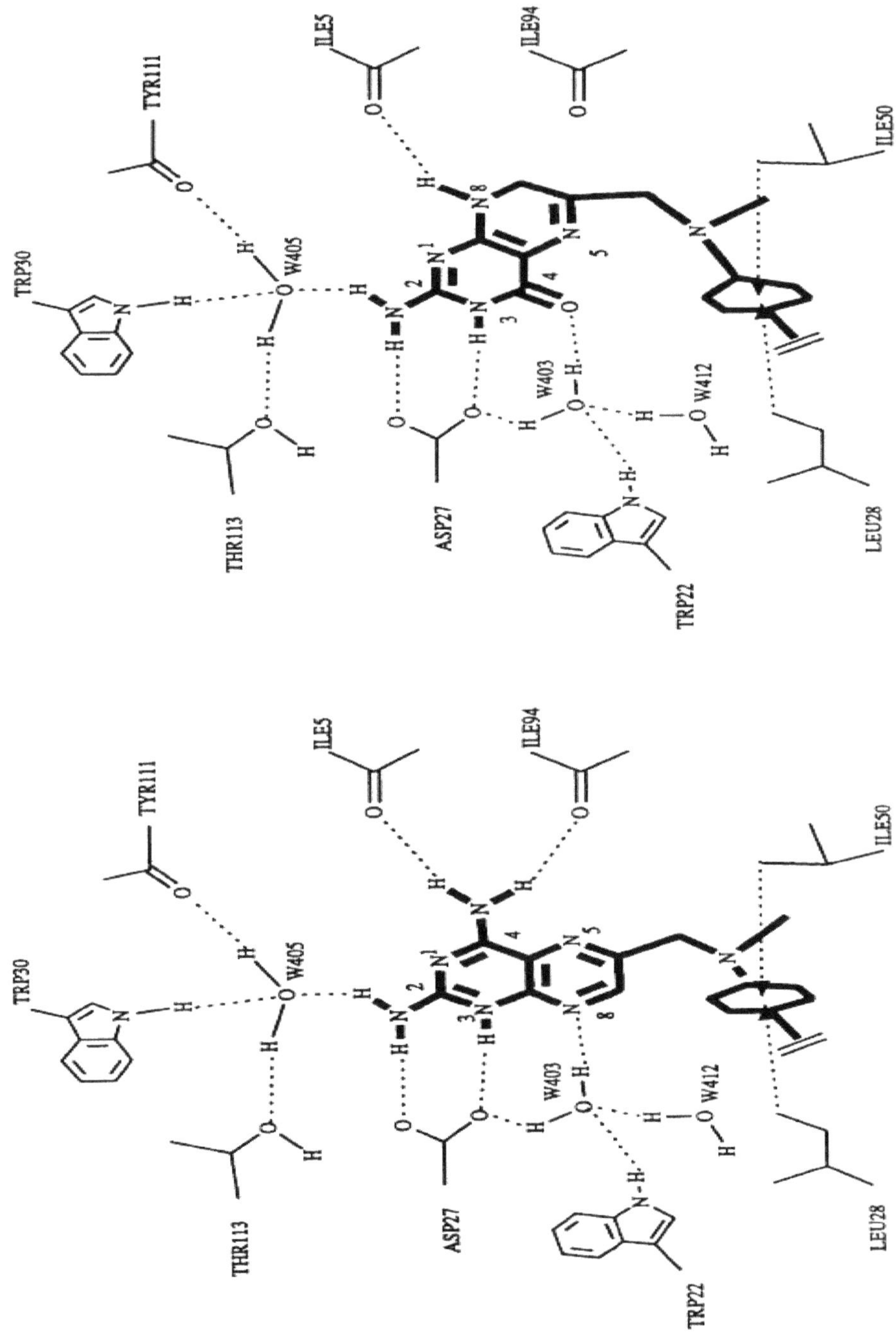

Figure 9.16 Interactions of the dihydrofolate reductase active site with the inhibitor methotrexate (left) and the substrate dihydrofolate (right).

With the modeled or actual structure of the target enzyme active site in hand, the next step is to assess the active site structure in a meaningful way, to permit the use of this information to predict inhibitor binding motifs. The simple visual inspection of such structures can be augmented today with computer programs that allow the analyst to map the electrostatic potential surface of the active site, identify and localize specific types of functional group within the active site (potential acid–base groups, hydrogen-bonding acceptors or donors, etc.), and the like. When the active site has been well described, one attempts to design inhibitors with stereochemical and functional complementarity to the active site structure. Again, these activities are greatly aided by high powered computer programs that make possible the probing of complementarity between a potential inhibitor and the enzyme active site. Assessment of the stereochemical complementarity of a potential inhibitor is aided by the use of molecular dynamics simulation programs by means of which the most energetically favorable conformations of inhibitory molecules can be assessed to determine whether they will adapt a conformation that is complementary to the enzyme active site.

New programs allow one to perform free energy perturbation calculations in which a bound inhibitor is slowly mutated and the difference in calculated free energy of binding between the starting and final structures determined . In this way, one can search for structural perturbations that will increase the affinity of an inhibitor for the enzyme active site. The complementarity of functional groups can be probed by computational methods as well. For example, the computer program GRID can be used to search the structure of an enzyme active site for areas that are likely to interact strongly with a particular functional group probe.

A recent example of the use of such programs comes from the studies by von Itzstein et al. aimed at designing potent inhibitors of the sialidase enzyme from influenza virus. This group started with the enzyme active site obtained from a series of crystals for the enzyme to which various sialic acid analogues were bound. Visual inspection of the cocrystal structure for the enzyme bound to the unsaturated sialic acid analogue Neu5Ac2en suggested that replacement of the 4-hydroxyl group of the substrate by an amino group might be useful. A GRID calculation was performed with a protonated primary amine group as the probe, and a "hot spot," or area of likely strong interaction, was identified within the enzyme active site.

The results of this process of visual inspection and calculation suggested that replacement of the 4-hydroxyl group with an amino group would lead to much tighter binding because a salt bridge would form between the amino group and the side chain carboxylate of Glu 119 of the enzyme. Further evaluation of the computational data suggested that replacement of the 4-hydroxyl group with a guanidinyl group would even further enhance inhibitor binding by engaging both Glu 119 and Glu 227through lateral binding of the two terminal nitrogens on this functional group. Based on these results, the 4-amino and 4-guanidino derivatives of Neu5Ac2en were synthesized and, as expected, found to be potent inhibitors of the enzyme, with K_m values of 50 and 0.2 nM, respectively. When the crystal structures of the enzyme bound to each of these new inhibitors was determined, the predicted modes of inhibitor interactions with the enzyme were by and large confirmed.

Chapter 10

TIGHT BINDING INHIBITORS

We are able to analyze the interactions of these inhibitors with their target enzymes by means of equations of the Henri–Michaelis–Menten type discussed in Chapters 5 and 9 because we can generally assume that the free inhibitor concentration is well modeled by the total concentration of added inhibitor:that is, since [E] is much smaller than K_i, the concentration of the EI complex is held to be very small compared to [I]. This assumption, however, is not valid for all inhibitors. Some inhibitors bind to their target enzyme with such high affinity that the population of free inhibitor molecules is significantly depleted by formation of the enzymeinhibitor complex. For these *tight binding inhibitors*, the steady state approximations used thus far are no longer valid; in fact, it has been suggested that these assumptions should be abandoned whenever the K_i of an inhibitor is less than 1000-fold greater than the total enzyme concentration . In this chapter we shall describe alternative methods for data analysis in the case of tight binding inhibitors that allow us to characterize the type of inhibition mechanism involved and to quantify correctly the dissociation constant for the enzyme–inhibitor complex.

10.1 IDENTIFYING TIGHT BINDING INHIBITION

In this chapter we shall consider the steady state approach to studying tight binding inhibitors. Such work requires assay conditions that permit all the equilibria involving the inhibitor, substrate, and enzyme to be reached rapidly with respect to the measurement of the steady state velocities. Many tight binding inhibitors, however, are restricted in their action by a slow onset of inhibition—that is, by a time-dependent component to their inhibition. We deal with time-dependent inhibition explicitly in another Chapter . For our present discussion, we shall assume either that the establishment of equilibrium is rapid or that sufficient time has been allowed before the initiation of reaction by the substrate for the inhibitor and enzyme to establish equilibrium (i.e., a preincubation of the enzyme with the inhibitor has been incorporated into the experimental design).

The simplest determination that tight binding inhibition is occurring comes from measurement of the dose–response curve for inhibition . An IC_{50} value obtained from this treatment of the data that is similar to the concentration of total enzyme in the sample (i.e., within a factor of 10) is a good indication that the inhibitor is of the tight binding type. A more defining feature of tight binding inhibitors is the variation of the IC_{50} value observed for these inhibitors with total enzyme concentration at a fixed substrate concentration. This is true because a tight binding inhibitor interacts with the enzyme in nearly stoichiometric fashion. Hence, the higher the concentration of enzyme present, the higher the concentration of inhibitor required to reach half-maximal saturation of the inhibitor binding sites (Figure 10.1A). Several authors have derived equations similar in form to Equation 10.1, which demonstrates that the IC_{50} value of a tight binding inhibitor will track linearly with the total concentration of enzyme, [E] :

$$IC_{50} = \tfrac{1}{2}[E] + K_i^{app} \qquad (10.1)$$

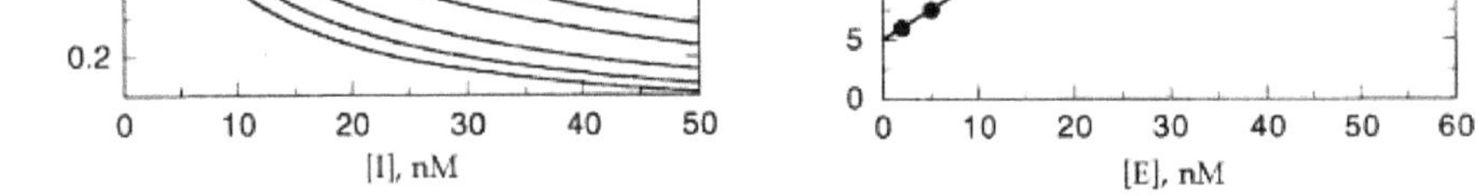

Figure 10.1 (A) Dose–response plot of fractional velocity as a function of tight binding inhibitor concentration at different enzyme concentrations. Note that this plot is the same as the dose–response plots introduced in Chapter 8, except that here the x axis is plotted on a linear, rather than a logarithmic, scale. (B) Plot of IC_{50} value obtained from the curves in (A) as afunction of enzyme concentration.

Thus, a plot of IC_{50} as a function of [E] (at a single, fixed substrateconcentration) is expected to yield a straight line with slope of 0.5 and *y*intercept equal to K_i^{apr}. The value K_i^{apr} is related to the true K_i by factorsinvolving the substrate concentration and K_m, depending on the mode ofinteraction between the inhibitor and the enzyme.

10.2 DISTINGUISHING INHIBITOR TYPE FOR TIGHT BINDING INHIBITORS

Morrison has provided indepth mathematical treatments of the effects of tight binding inhibitors on the initial velocities of enzymatic reactions. These studies revealed, among other things, that the classical double-reciprocal plots used to distinguish inhibitor type for simple enzyme inhibitors fail in the case of tight binding inhibitors.

For example, based on the work just cited by Morrison and coworkers, the double-reciprocal plot for a tight binding competitive inhibitor would give the pattern of lines illustrated in Figure 10.2. The data at very high substrate concentrations curve downward in this plot, and the curves at different inhibitor concentrations converge at the *y* axis. Note, however, that this curvature is apparent only at very high substrate concentrations and in the presence of high inhibitor concentrations. This subtlety in the data analysis is easy to miss if care is not taken to include such extreme conditions, or if these conditions are not experimentally attainable. Hence, if the few data points in the very high substrate region are ignored, it is tempting to fit the data in Figure 10.2 to a series of linear functions, as has been done in this illustration.

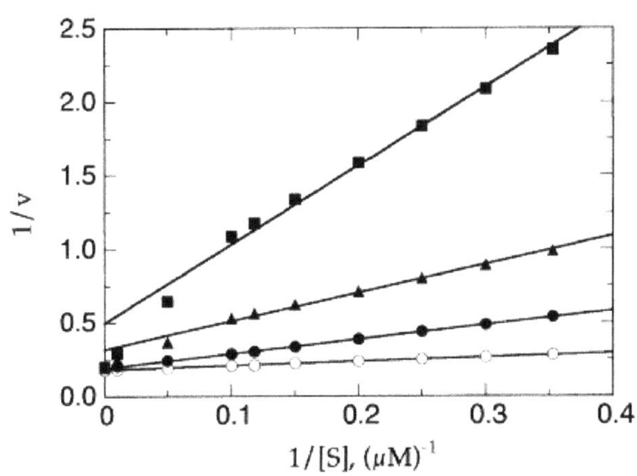

Figure 10.2 Double-reciprocal plot for a tight binding competitive inhibitor: the pattern of lines is similar to that expected for a classical noncompetitive inhibitor.

The pattern of lines that emerges from this treatment of the data is a series of lines that intersect at or near the *x* axis, to the left of the *y* axis. This is the expected result for a classical noncompetitive inhibitor, and we can generally state that *regardless of their true mode of interaction with the enzyme, tight binding inhibitors display double-reciprocal plots that appear similar to the classical pattern for noncompetitive inhibitors.*

As one might imagine, this point has led to a number of misinterpretations of kinetic data for inhibitors in the literature. For example, the naturally occurring inhibitors of ribonuclease are nanomolar inhibitors of the enzyme. Initial evaluation of the inhibitor type by double-reciprocal plots indicated that these inhibitors acted through classical noncompetitive inhibition. It was not until Turner et al. performed a careful examination of these inhibitors, over a broad range of inhibitor and substrate concentrations, and properly evaluated the data (as discussed below) that these proteins were recognized to be tight binding *competitive* inhibitors.

How then can one determine the true mode of interaction between an enzyme and a tight binding inhibitor? Several graphical approaches have been suggested. One of the most straightforward is to determine the IC_{50} values for the inhibitor at a fixed enzyme concentration, but at a number of different substrate concentrations. As with simple reversible inhibitors, the IC_{50} of a tight binding inhibitor depends on the K_i of the inhibitor, the substrate concentration, and the substrate K_m in different ways, depending on the mode of inhibition. For tight binding inhibitors we must additionally take into consideration the enzyme concentration in the sample, since this will affect the measured IC_{50}, as discussed earlier. The appropriate relationships betweenthese factors and the IC_{50} for different types of tight binding inhibitor havebeen derived several times in the literature . Rather than working through these derivationsagain, we shall simply present the final form of the relationships. For tight binding competitive inhibitors:

$$IC_{50} = K_i\left(1 + \frac{[S]}{K_m}\right) + \frac{[E]}{2} \tag{10.2}$$

For tight binding noncompetitive inhibitors:

$$IC_{50} = \frac{[S] + K_m}{\dfrac{K_m}{K_i} + \dfrac{[S]}{\alpha K_i}} + \frac{[E]}{2} \tag{10.3}$$

when α=1:

$$IC_{50} = K_i + \frac{[E]}{2} \tag{10.4}$$

For tight binding uncompetitive inhibitors:

$$IC_{50} = \alpha K_i \left(1 + \frac{K_m}{[S]}\right) + \frac{[E]}{2} \tag{10.5}$$

From the form of these equations, we see that a plot of the IC_{50} value as a function of substrate concentration will yield quite different patterns, depending on the inhibitor type. For a tight binding competitive inhibitor, the IC50 value will increase linearly with increasing substrate concentration (Figure10.3A). For an uncompetitive inhibitor, a plot of IC_{50} value as a function of substrate concentration will curve downward sharply (Figure 10.3A), while fora noncompetitive inhibitor the IC_{50} will curve upward or downward, or be independent of [S], depending on whether α is greater than, less than, or equal to 1.0 (Figure 10.3A and B).

In an alternative graphical method for determining the inhibitor type, and obtaining an estimate of the inhibitor K_i , the fractional velocity of the enzyme reaction is plotted as a function of inhibitor concentration at some fixed substrate concentratio. The data can be fit to Equation 10.20 to yield a curvilinear fit as shown in Figure 10.4A. (Note that this is the same as the dose—response plots discussed in earlier Chapter , except here the *x* axis is plotted on a linear, rather than a logarithmic, scale). A line is drawn from thev/v_0 value at [I]_0 (referred to here as the starting point) through the point on the curve where $v=v_0/2$ (n=2) and extended to the *x* axis. A second lineis drawn from the starting point through the point on the curve where $v=v_0/3$(n=3), and, in a similar fashion, additional lines are drawn from the startingpoint through other points on the curve where $v=v_0/n$ (where *n* is an integer).

The nest of lines thus drawn will intersect the *x* axis at a constant spacing, which is defined as *K*.

Knowing the value of *K* from a nest of these lines, one can draw additional lines from the *x* axis to the origin at spacing of *K* on the *x* axis, for apparent values of *n*=1 and *n*=0. From this treatment, the line corresponding to *n*=0 will intersect the *x* axis at a displacement from the origin that is equal to the total enzyme concentration, [E]. Dixon goes on to show that in the case of a noncompetitive inhibitor (α=1), the spacing value *K* is equal to the inhibitor K_i , and a plot of *K* as a function of substrate concentration will be a horizontal line; that is, the value of *K* for a noncompetitive inhibitor is independent of substrate concentration. For a competitive inhibitor, however, the measured value of *K* will increase with increasing substrate concentration. A replot of K as a function of substrate concentration yields estimates of the K_i of the inhibitor and the K_m of the substrate from the y and x intercepts, respectively (Figure 10.4B).

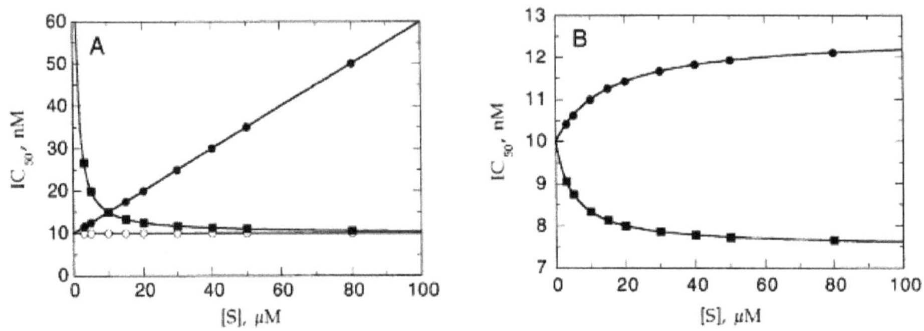

Figure 10.3 (A) The effects of substrate concentration on the IC_{50}values of competitive (solidcircles), noncompetitive when α=1 (open circles), and uncompetitive (solid squares) tightbinding inhibitors. (B) The effects of substrate concentration on the IC_{50}values of noncompetitivetight binding inhibitors when α<1 (squares) and when α>1 (circles).

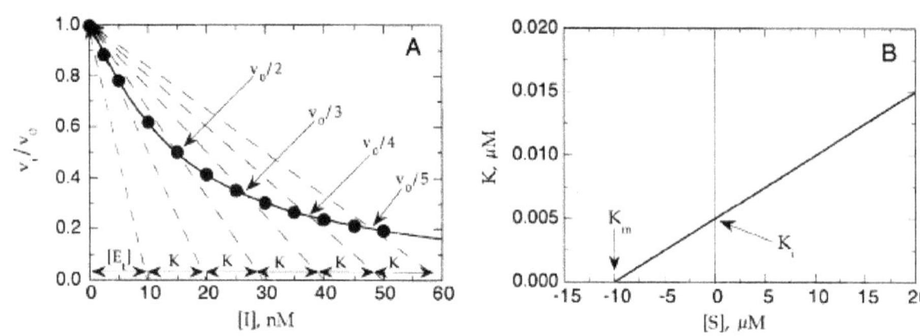

Figure 10.4 (A) Determination of "K" by the graphical method of Dixon : dashed lines connect the starting point (v_i/v_0=1, [I]=0) with points on the curve where $v_i/v_0=v_0/n$ (n=2, 3, 4, and 5). Additional lines are drawn for apparent n=1 and apparent n=0, based on thex-axis spacing value "K," determine from the n=2—5 lines (see text for further details). (B)Secondary plot of the "K" as a function of substrate concentration for a tight binding competitiveinhibitor. Graphical determinations of K_i and K_m are obtained from the values of the y and xintercepts of the plot, respectively, as shown.

10.3 DETERMINING KI FOR TIGHT BINDING INHIBITORS

The literature describes several methods for determining the K_i value of a tight binding enzyme inhibitor. We have already discussed the graphical method of Dixon, which allows one to simultaneously distinguish inhibitor type and calculate the K_i. A more mathematical treatment of tight binding inhibitors, presented by Morrison, led to a generalized equation to describe the fractional velocity of an enzymatic reaction as a function of inhibitor concentration, at fixed concentrations of enzyme and substrate. This equation, commonly referred to as the Morrison equation, is derived in a manner similar to Equation 4.38, except that here the equation is cast in terms of fractional enzymatic activity in the presence of the inhibitor (i.e., in terms of the fraction of free enzyme instead of the fraction of inhibitor-bound enzyme).

$$\frac{v_i}{v_0} = 1 - \frac{([E] + [I] + K_i^{app}) - \sqrt{([E] + [I] + K_i^{app})^2 - 4[E][I]}}{2[E]} \quad (10.6)$$

The form of K_i^{app} in Equation 10.6 varies with inhibitor type. The following explicit forms of this parameter for the different inhibitor types are similar to those presented in Equations 10.2—10.5 for the IC_{50} values.

For competitive inhibitors:

$$K_i^{app} = K_i\left(1 + \frac{[S]}{K_m}\right) \quad (10.7)$$

For Noncompetitive Inhibitors:

$$K_i^{app} = \frac{[S] + K_m}{\frac{K_m}{K_i} + \frac{[S]}{\alpha K_i}} \quad (10.8)$$

when α=1:

$$K_i^{app} = K_i \quad (10.9)$$

For uncompetitive inhibitors:

$$K_i^{app} = K_i\left(1 + \frac{K_m}{[S]}\right) \quad (10.10)$$

Prior to the widespread use of computer-based routines for curve fitting, the direct use of the Morrison equation was inconvenient for extracting inhibitor constants from experimental data. To overcome this limitation, Henderson presented the derivation of a linearized form of the Morrison equation that allowed graphical determination of K_i and [E] from measurements of the fractional velocity as a function of inhibitor concentration at a fixed substrate concentration. The generalized form of the Henderson equation is as follows:

$$\frac{[\mathrm{I}]}{1-\frac{v_i}{v_0}} = K_i^{app}\left(\frac{v_0}{v_i}\right) + [\mathrm{E}] \tag{10.11}$$

where K_i^{app} has the same forms as presented in Equations 10.7—10.10 for the various inhibitor types.

Inspection reveals that Equation 10.11 is a linear equation. Hence, if one were to plot $[I]/(1-v_i/v_0)$ as a function of v_0/v_i (i.e., the reciprocal of the fractional velocity), the data could be fit to a straight line with slope equal to K_i^{app} and y intercept equal to [E], as illustrated in Figure 10.5. Note that the Henderson method yields a straight-line plot regardless of the inhibitor type. The slope of the lines for such plots will, however, vary with substrate concentration in different ways depending on the inhibitor type. The variation observed is similar to that presented in Figure 10.3 for the variation in IC_{50} value for different tight binding inhibitors as a function of substrate concentration. Thus, the Henderson plots also can be used to distinguish among the varying inhibitor binding mechanisms.

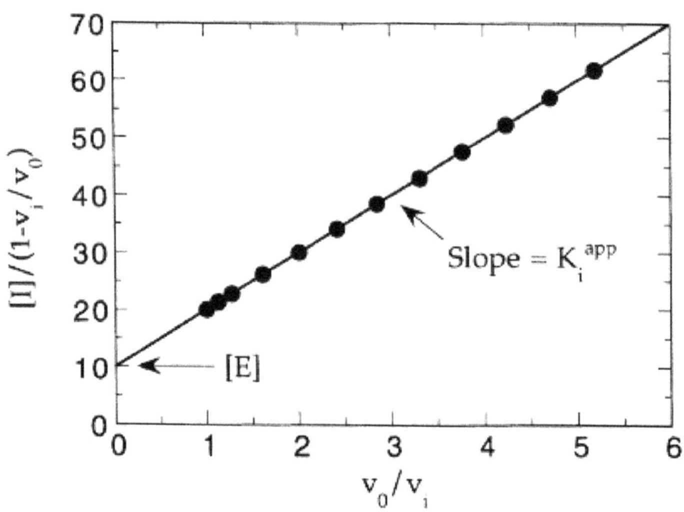

Figure 10.5 Henderson plot for a tight binding inhibitor.

While linearized Henderson plots are convenient in the absence of a computer curve-fitting program, the data treatment does introduce some degree of systematic error . Today, with the availability of robust curve-fitting routines on laboratory computers, it is no longer necessary to resort to linearized treatments of data such as the Henderson plots. The direct fitting of fraction velocity versus inhibitor concentration data to the Morrison equation (Equation 10.6) is thus much more desirable, and is strongly recommended. Figure 10.6 illustrates the direct fitting of fractional velocity versus inhibitor concentration data to Equation 10.6. Such data would call for predetermination of the K_m value for the substrate and knowledge of the substrate concentration in the assays. Then the data, such as the points in Figure 10.6, would be fit to the Morrison equation, allowing both K_i^{app} and [E] to be simultaneously determined as fitting parameters. Measurements of this type at several different substrate concentrations would allow determination of the mode of inhibition, and thus the experimentally measured K_i^{app} values could be converted to true K_i values.

In the case of competitive tight binding inhibitors, an alternative method for determining inhibitor K_i is to measure the iniital velocity under conditions of extremely high substrate concentration . Reflecting on Equation 10.2, we see that if the ratio $[S]/K_m$ is very large, the IC_{50} Will be much greater than the enzyme concentration, even though the K_i is similar in magnitude to [E]. Thus, if a high enough substrate concentration can be experimentally achieved, the tight binding nature of the inhibitor can be overcome, and the K_i can be determined from the measured IC_{50} by application of a rearranged form of Equation 10.2. Tornheim recommends adjusting [S] so that the ratios $[S]/K_m$ and $[I]/K_i$ are about equal for these measurements. Not all enzymatic reactions are amenable to this approach, however, because of the experimental limitations on substrate concentration imposed by the solubility of the substrate and the analyst's ability to measure a linear initial velocity under such extreme conditions. In favorable cases, however, this approach can be used with excellent results.

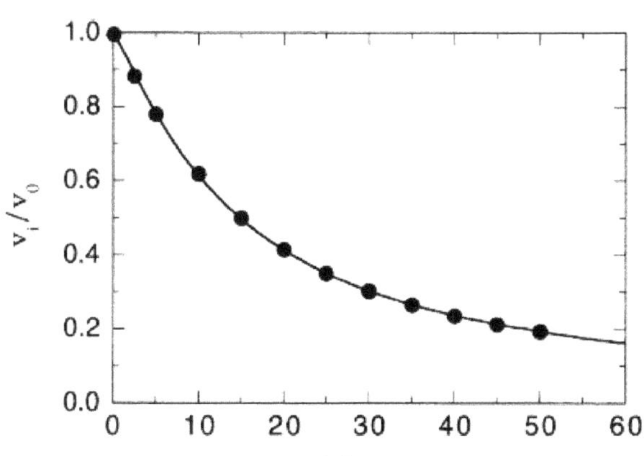

Figure 10.6 Plot of fractional velocity as a function of inhibitor concentration for a tight binding inhibitor. The solid curve drawn through the data points represents the best fit to the Morrison equation (Equation 10.6).

10.4 USE OF TIGHT BINDING INHIBITORS TO DETERMINE ACTIVE ENZYME CONCENTRATION

In many experimental strategies one wishes to know the concentration of enzyme in a sample for subsequent data analysis. This approach applies not only to kinetic data, but also to other types of biochemical and biophysical studies with enzymes. The literature gives numerous methods for determining total protein concentration in a sample, on the basis of spectroscopic, colorimetric, and other analytical techniques. All these methods, however, measure bulk protein concentration rather than the concentration of the target enzyme in particular. Also, these methods do not necessarily distinguish between active enzyme molecules, and molecules of denatured enzyme. In many of the applications one is likely to encounter, it is the concentration of active enzyme molecules that is most relevant. The availability of a tight binding inhibitor of the target enzyme provides a convenient means of accurately determining the concentration of active enzyme in the sample, even in the presence of denatured enzyme or other nonenzymatic proteins.

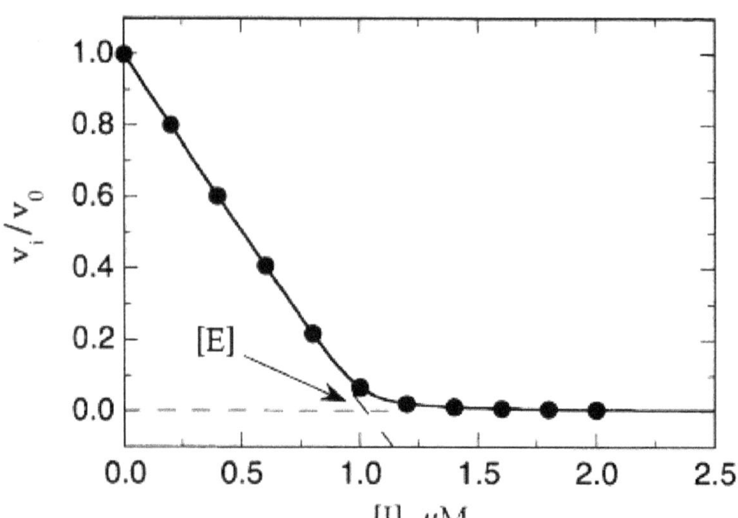

Figure 10.7 Determination of active enzyme concentration by titration with a tight binding inhibitor. [E]=1.0 μM, K_i=5 nM (i.e., [E]/K_i=200). The solid curve drawn through the data is the best fit to the Morrison equation (Equation 10.6). The dashed lines were drawn by linear least-squares fits of the data at inhibitor concentrations that were low (0—0.6 μM) and high (1.4—2.0 μM), respectively. The active enzyme concentration is determined from the x-axis value at the intersection of the two straight lines.

Referring back to Equation 10.6, if we set up an experiment in which both[E] and [I] are much greater than K_i^{app}, we can largely ignore the K_i^{app} term in this equation. Under these conditions, the fractional velocity of the enzymatic reaction will fall off quasi-linearly with increasing inhibitor concentration until [I]=[E]. At this point the fractional velocity will approach zero and remain there at higher inhibitor concentrations. In this case, a plot of fractional velocity as a function of inhibitor concentration will look like Figure 10.7 when fit to the Morrison equation. The data in figure 10.7 were generated for a hypothetical situation: *K*_ of inhibitor, 5 nM; active enzyme concentration of the sample, 1.0 μM (i.e., $[E]/K_i$=200). The data at lower inhibitor concentration can be fit to a straight line that is extended to the *x* axis (dashed line in Figure 10.7), and the data points at higher inhibitor concentrations can be fit to a straight horizontal line at v_i/v_0=0 (longer dashed line in Figure 10.7). The two lines thus drawn will intersect at a point on the *x* axis where [I]=[E]. bNote, however, that this treatment works only when [E] is much greater than K_i. When [E] is less than about 200K_i, the data are not well described by two intersecting straight lines. In such cases the data can be fit directly to Equation 10.6 to determine [E], as described earlier.

This type of treatment is quite convenient for determining the active enzyme concentration of a stock enzyme solution (i.e., at high enzyme concentration) that will be diluted into a final reaction mixture for experimentation. For example, one might wish to store an enzyme sample at a nominal enzyme concentration of 100 μM in a solution containing 1 mg/mL gelatin for stability purposes . The presence of the gelatin would preclude accurate determination of enzyme concentration by one of the traditional colorimetric protein assays; moreover, *active* enzyme concentration could not be determined by means of such assays. Given a nanomolar inhibitor of the target enzyme, one could dilute a sample of the stock enzyme to some convenient concentration for an enzymatic assay that was still much greater than the K_i(e.g., 1 μM). Treatment of the fractional velocity versus inhibitor concentration as described here would thus lead to determination of the true concentration of active enzyme in the working solution, and from this one could back-calculate to arrive at the true concentration of active enzyme in the enzyme stock. This is a routine strategy in many enzymology laboratories, and numerous examples of its application can be found in the literature.

A comparable assessment of active enzyme concentration can be obtained by the reverse experiment in which the inhibitor concentration is fixed at some value much greater than the K_i (about 200 K_i or more), and the amount of enzyme added to the reaction mixture is varied. The results of such an experiment are illustrated in Figure 10.8. The initial velocity remains zero until equal concentrations of enzyme and inhibitor are present in solution. As the enzyme concentration is titrated beyond this point, the stoichiometric inhibition is overcome, and a linear increase in initial velocity is then observed.

Again, from the point of intersection of the two dashed lines drawn through the data as in Figure 10-.8, the true concentration of active enzyme can be determined . An advantage of this second approach to active enzyme concentration determination is that it typically uses up less of the enzyme stock to complete the titration. Hence, when the enzyme is in limited supply, this alternative is recommended.

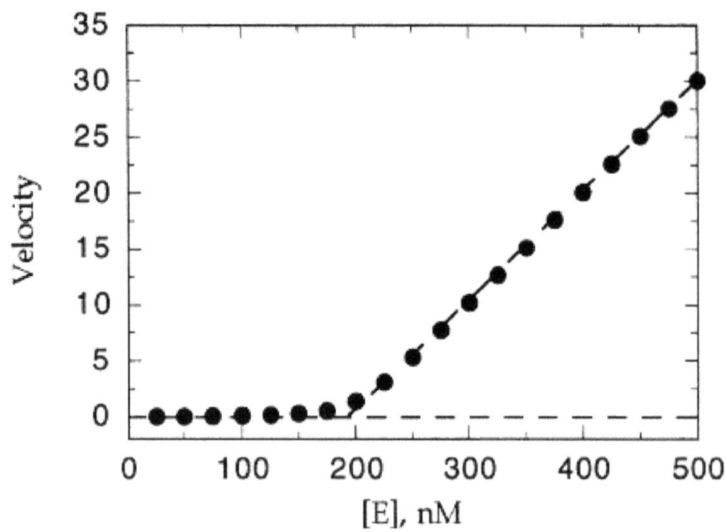

Figure 10.8 Determination of active enzyme concentration by titration of a fixed concentration of a tight binding inhibitor with enzyme: [I]=200 nM, K_i =1 nM (i.e., [I]/K_i =200). The data analysis is similar to that described for Figure 10.7 and in the text. Velocity is in arbitrary units.

Chapter 11

ENZYME REACTIONS WITH MULTIPLE SUBSTRATES

Until now we have considered only the simplest of enzymatic reactions, those involving a single substrate being transformed into a single product. However, the vast majority of enzymatic reactions one is likely to encounter involve at least two substrates and result in the formation of more than one product. Let us look back at some of the enzymatic reactions we have used as examples. Many of them are multisubstrate and/or multiproduct reactions. For example, the serine proteases selected to illustrate different concepts in earlier chapters use two substrates to form two products. The first, and most obvious, substrate is the peptide that is hydrolyzed to form the two peptide fragment products. The second, less obvious, substrate is a water molecule that indirectly supplies the proton and hydroxyl groups required to complete the hydrolysis. Likewise, when we discussed the phosphorylation of proteins by kinases, we needed a source of phosphate for the reaction, and this phosphate source itself is a substrate of the enzyme. An ATP-dependent kinase, for example, requires the protein and ATP as its two substrates, and it yields the phosphoprotein and ADP as the two products. A bit of reflection will show that many of the enzymatic reactions in biochemistry proceed with the use of multiple substrates and/or produce multiple products. In this chapter we explicitly deal with the steady state kinetic approach to studying enzyme reactions of this type.

11.1 REACTION NOMENCLATURE

A general nomenclature has been devised to describe the number of substrates and products involved in an enzymatic reaction, using the Latin prefixes uni, bi, ter, and so on to refer to one, two, three, and more chemical entities. For example, a reaction that utilizes two substrates to produce two products is referred to as a *bi bi* reaction, a reaction using three substrates to form two products is as a *ter bi* reaction, and so on (Table 11.1).

Table 11.1 General nomenclature for enzymatic Reactions

Reaction	Name
$A \longrightarrow P$	Uni uni
$A + B \longrightarrow P$	Bi uni
$A + B \longrightarrow P_1 + P_2$	Bi bi
$A + B + C \longrightarrow P_1 + P_2$	Ter bi
⋮	⋮

Let us consider in some detail a group transfer reaction that proceeds as a bi bi reaction:

$$E + AX + B \rightleftharpoons E + A + BX$$

The reaction scheme as written leaves several important questions unanswered.

Does one substrate bind and leave before the second substrate can bind? Is the order in which the substrates bind random, or must binding occur in a specific sequence? Does group X transfer directly from A to B when both are bound at the active site of the enzyme, or does the reaction proceed by transfer of the group from the donor molecule, A, to a site on the enzyme, whereupon there is a second transfer of the group from the enzyme site to the acceptor molecule B (i.e., a reaction that proceeds through formation of an E—X intermediate)?

These questions raise the potential for at least three distinct mechanisms for the generalized scheme; these are referred to as random ordered, compulsory ordered, and double-displacement or "Ping-Pong" bi bi mechanisms. Often a major goal of steady state kinetic measurements is to differentiate between these varied mechanisms. We shall therefore present a description of each and describe graphical methods for distinguishing among them.

In the treatments that follow we shall use the general steady state rate equations of Alberty (1953), which cast multisubstrate reactions in terms of the equilibrium constants that are familiar from our discussions of the Henri– Michaelis–Menten equation. This approach works well for enzymes that utilize one or two substrates and produce one or two products.

11.2 BI BI REACTION MECHANISMS

11.2.1 Random Ordered Bi Bi Reactions

In the random ordered bi bi mechanism, either substrate can bind first to the enzyme, and either product can leave first. Regardless of which substrate binds first, the reaction goes through an intermediate ternary complex (E·AX· B), as illustrated:

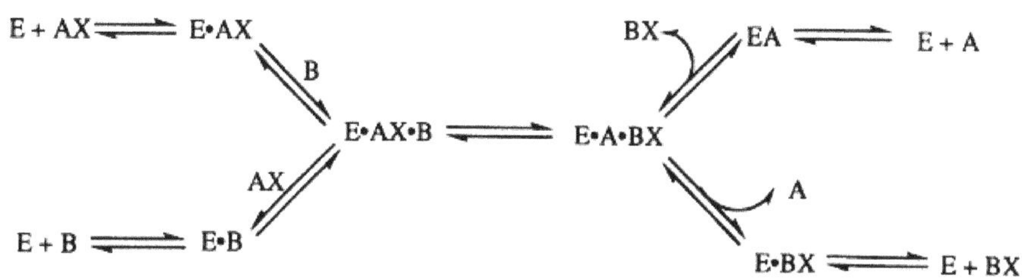

Here the binding of AX to the free enzyme (E) is described by the dissociation constant K^{AX}, and the binding of B to E is likewise described by K^{B}. Note that the binding of one substrate may very well affect the affinity of the enzyme for the second substrate. Hence, we may find that the binding of AX to the preformed E ·B complex is described by the constant αK^{AX}. Likewise, since the overall equilibrium between E·AX·B and E must be path independent, the binding of B to the preformed E ·AX complex is described by αK^{B}. When B is saturating, the value of αK^{AX} is equal to the Michaelis constant for AX (i.e., K_m^{AX}). Likewise, when AX is saturating, $\alpha K^{B}=K_m^{B}$. The velocity of such an enzymatic reaction is given by Equation 11.1:

$$v = k_{cat}[E \cdot AX \cdot B] = \frac{k_{cat}[E_t][E \cdot AX \cdot B]}{[E] + [E \cdot AX] + [E \cdot B] + [E \cdot AX \cdot B]} \quad (11.1)$$

If we express the concentrations of the various species in terms of the free enzyme concentration [E], we obtain:

$$v = \frac{V_{max}[AX][B]}{\alpha K^{AX}K^{B} + \alpha K^{B}[AX] + \alpha K^{AX}[B] + [AX][B]} \quad (11.2)$$

If we fix the concentration of one of the substrates, we can rearrange and simplify Equation 11.2 significantly. For example, when [B] is fixed and [AX] varies, we obtain:

$$v = \frac{V_{max}[AX]}{\alpha K^{AX}\left(1 + \frac{K^{B}}{[B]}\right) + [AX]\left(1 + \frac{\alpha K^{B}}{[B]}\right)}$$

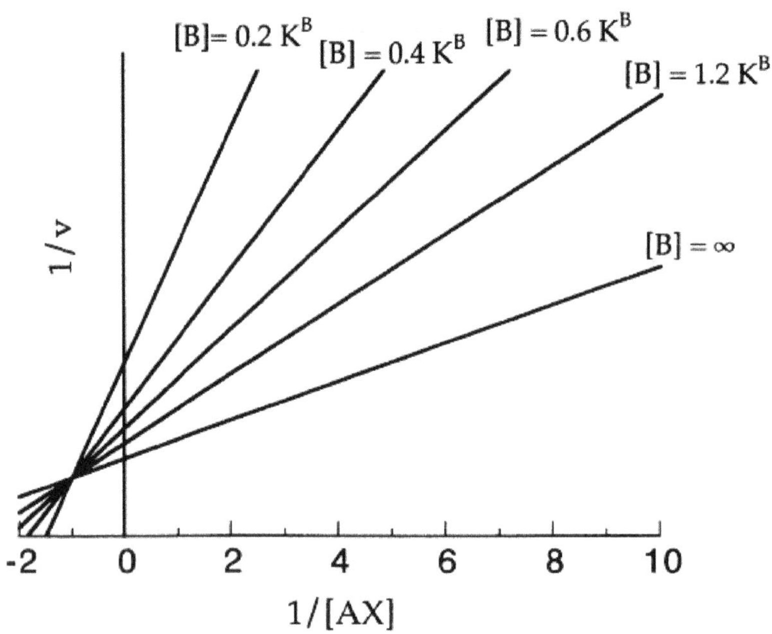

Figure 11.1 Double-reciprocal plot for a random ordered bi bi enzymatic reaction.

At high, fixed concentrations of B, the terms $K^{B}/[B]$ and $\alpha K^{B}/[B]$ go to zero.

Thus, at saturating concentrations of B we find:

$$v = \frac{V_{max}^{app}[AX]}{K_{m}^{AX,app} + [AX]} \tag{11.4}$$

and likewise, at fixed, saturating [AX]:

$$v = \frac{V_{max}^{app}[B]}{K_{m}^{B,app} + [B]} \tag{11.5}$$

If we measure the reaction velocity over a range of AX concentrations at several, fixed concentrations of B, the reciprocal plots will display a nest of lines that converge to the left of the *y* axis, as illustrated in Figure 11.1. The data from Figure 11.1 can be replotted as the slopes of the lines as a function of 1/[B], and the *y* intercepts (i.e., $1/V_{max}^{app}$) as a function of 1/[B] (Figure 11.2).

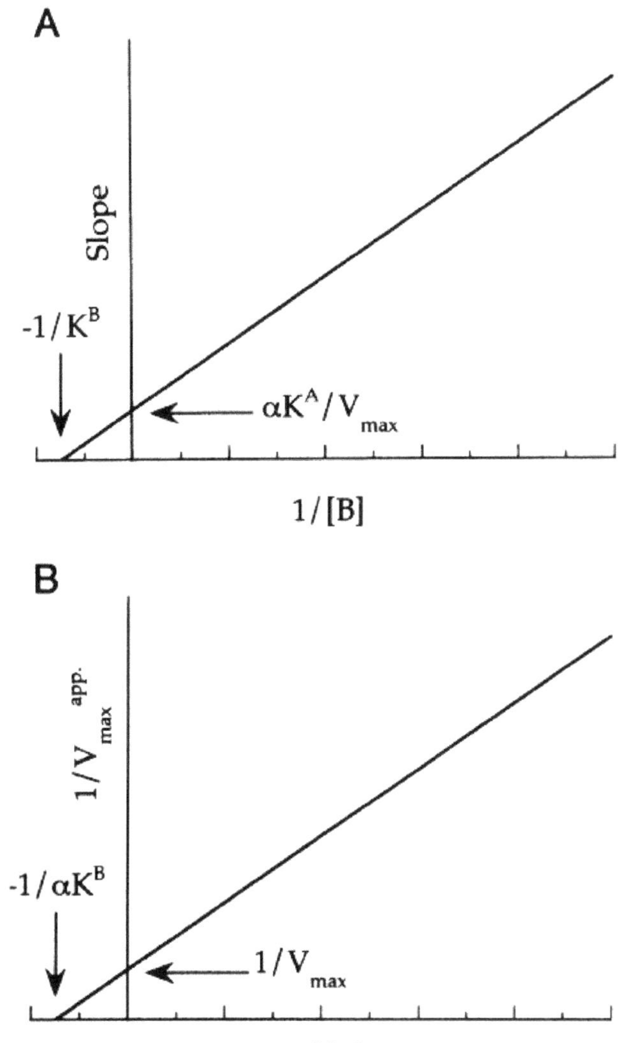

Figure 11.2 (A) Slope and (B) y-intercept replots of the data from Figure 11.1, illustrating the graphical determination of K^{AX}, K^{B}, and V_{max} for a random ordered bi bi enzymatic reaction.

11.2.2 Compulsory Ordered Bi Bi Reactions

In compulsory ordered bi bi reactions, one substrate, say AX, must bind to the enzyme before the other substrate (B) can bind. As with random ordered reactions, the mechanism proceeds through formation of a ternary intermediate. In this case the reaction scheme is as follows:

$$\mathrm{E + AX} \rightleftharpoons \mathrm{E \cdot AX} \overset{\mathrm{B}}{\rightleftharpoons} \mathrm{E \cdot AX \cdot B} \rightleftharpoons \mathrm{E \cdot A \cdot BX} \overset{-\mathrm{BX}}{\rightleftharpoons} \mathrm{E \cdot A} \rightleftharpoons \mathrm{E + A}$$

If conversion of the E ·AX·B complex to E ·A·BX is the rate-limiting step in catalysis, then E, AX, B, and E ·AX·B are all in equilibrium, and the velocity of the reaction will be given by:

$$v = \frac{V_{max}[\mathrm{AX}][\mathrm{B}]}{K^{\mathrm{AX}}K^{\mathrm{B}} + K^{\mathrm{B}}[\mathrm{AX}] + [\mathrm{AX}][\mathrm{B}]} \tag{11.6}$$

If, however, the conversion of E·AX·B to E ·A·BX is as rapid as the other steps in catalysis, steady state assumptions must be used in the derivation of the velocity equation. For a compulsory ordered bi bi reaction, the steady state treatment yields Equation 11.7:

$$v = \frac{V_{max}[\mathrm{AX}][\mathrm{B}]}{K^{\mathrm{AX}}K_m^{\mathrm{B}} + K_m^{\mathrm{B}}[\mathrm{AX}] + K_m^{\mathrm{AX}}[\mathrm{B}] + [\mathrm{AX}][\mathrm{B}]} \tag{11.7}$$

As we have described before, the term K^{AX} in Equation 11.7 is the dissocation constant for the E ·AX complex, and K_m^{AX} is the concentration of AX that yields a velocity of half V_{max} at fixed, saturating [B].

The pattern of reciprocal plots observed for varied [AX] at different fixe values of [B] is identical to that seen in Figure 11.1 for a random ordered bi bi reaction (note the similarity between Equations 11.2 and 11.7). Hence, *one cannot distinguish between random and compulsory ordered bi bi mechanisms on the basis of reciprocal plots alone.* It is necessary to resort to the use of isotope incorporation studies, or studies using product-based inhibitors.

11.2.3 Double Displacement or Ping-Pong Bi Bi Reactions

The double displacement, or Ping-Pong, reaction mechanism involves binding of AX to the enzyme and transfer of the group, X, to some site on the enzyme. The product, A, can then leave and the second substrate, B, binds to the E—X form of the enzyme (in this mechanism, B cannot bind to the free enzyme form). The group, X, is then transferred (i.e., the second displacement reaction) to the bound substrate,

B, prior to the release from the enzyme of the final product, BX. This mechanism is diagrammed as follows:

$$E + AX \rightleftharpoons E \cdot AX \rightleftharpoons EX \cdot A \underset{}{\overset{-A}{\rightleftharpoons}} EX \overset{B}{\rightleftharpoons} EX \cdot B \rightleftharpoons E \cdot BX \rightleftharpoons E + BX$$

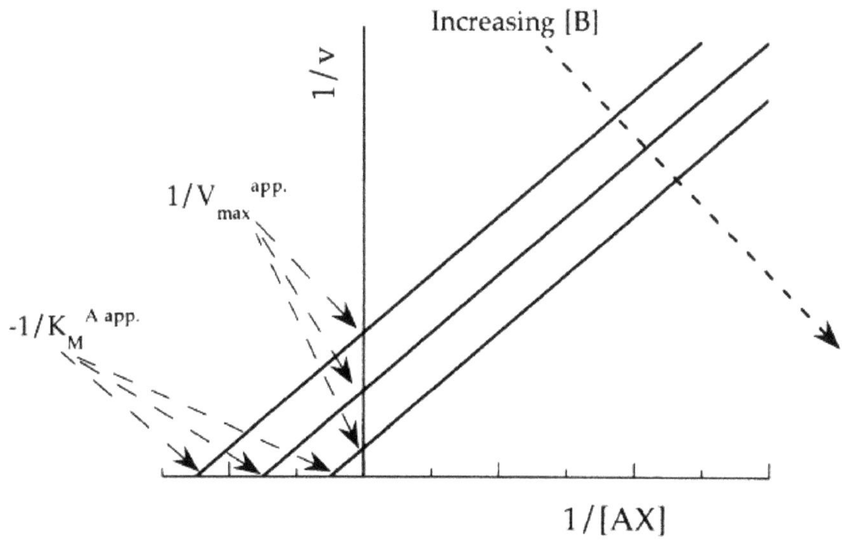

Figure 11.3 Double-reciprocal plot for a double-displacement (Ping-Pong) bi bi enzymatic reaction.

Using steady state assumptions, the velocity equation for a double-displacement reaction can be obtained:

$$v = \frac{V_{max}[AX][B]}{K_m^B[AX] + K_m^{AX}[B] + [AX][B]} \tag{11.8}$$

If we fix the value of [B], then Equation 11.8 for variable [AX] becomes:

$$v = \frac{V_{max}[AX]}{K_m^{AX} + [AX]\left(1 + \frac{K_m^B}{[B]}\right)} \tag{11.9}$$

Reciprocal plots of a reaction that conforms to the double-displacement mechanism for varying concentrations of AX at several fixed concentrations of B will yield a nest of parallel lines, as seen in Figure 11.3. For each concentration of substrate B, the values of $1/V_{max}^{app}$ and $-1/K_m^{AX,app}$ can be determined from the y and x intercepts, respectively, of the double-reciprocal plot. The data contained in Figure 11.3 can be replotted in terms of $1/V_{max}^{app}$ as a function of 1/[B], and $1/K_m^{AX,app}$ as illustrated in Figure 11.4. The value of $-1/K_m^B$ can be determined from the x intercepts of either replot in Figure 11.4. The y intercepts of the two replots yield estimates of $1/V_{max}$ (for the $1/V_{max}^{app}$ versus 1/[B] replot) and $1/K_m^{AX}$ (for the $1/K_m^{app}$ versus 1/[B] replot) for the reaction, as seen in Figure 11.4.

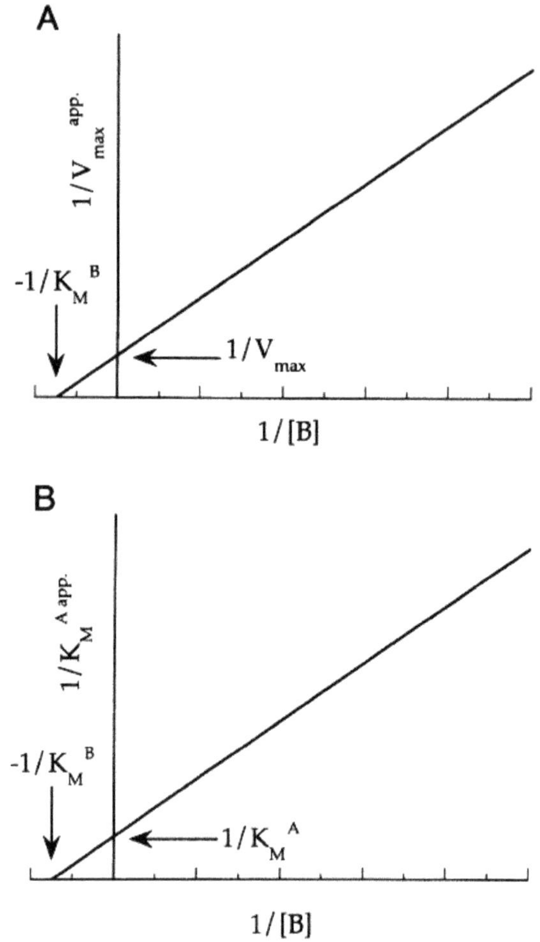

Figure 11.4 Replots of the data from Figure 11.3 as (A) $1/V_{max}^{app}$ versus 1/[B] and (B) $1/K_m^{AX,app}$ versus 1/[B], illustrating the graphical determination of K_m^{AX}, K_m^B, and V_{max} for a double-displacement (Ping-Pong) bi bi enzymatic reaction.

11.3 DISTINGUISHING BETWEEN RANDOM AND COMPULSORY ORDERED MECHANISMS BY INHIBITION PATTERN

It should be clear from Figures 11.1 and 11.3, and the foregoing discussion, that the qualitative form of the double-reciprocal plots makes it easy to distinguish between a double-displacement mechanism and a mechanism B=0 for the scheme) or product inhibitor, and the mechanism of substrate interaction with the enzyme. For a bi bi reaction, one observes specific inhibitor patterns for the different mechanisms we have discussed when a competitive dead-end inhibitor or a product of the reaction is used as the inhibitor. The patterns for both dead-end and product inhibition additionally depend on whether the fixed substrate is at a saturating or nonsaturating (typically at [S]~K_m) concentration with respect to its apparentK_m.

The relationships leading to these differing patterns of dead-end and product inhibition for bi bi reactions have been derived elsewhere.

Table 11.2 Patterns of dead-end inhibition observed for the Bi Bi reaction E+AX+B->E+A+BX for differing reaction mechanisms

Mechanism	Competitive Inhibitor for Substrate	Inhibitor Pattern Observed[a] For Varied [AX]	Inhibitor Pattern Observed[a] For Varied [B]
Compulsory ordered with [AX] binding first	AX	Competitive	Noncompetitive
Compulsory ordered with [AX] binding first	B	Uncompetitive	Competitive
Compulsory ordered with [B] binding first	AX	Competitive	Uncompetitive
Compulsory ordered with [B] binding first	B	Noncompetitive	Competitive
Random ordered	AX	Competitive	Noncompetitive
Random ordered	B	Noncompetitive	Competitive
Double displacement	AX	Competitive	Uncompetitive
Double displacement	B	Uncompetitive	Competitive

[a]At nonsaturating ([S] ~ K_m) concentration of the fixed substrate.

Table 11.3 Pattern of product inhibition observed for the Bi Bi reaction E+AX+B->E+A+BX for differing reaction mechanisms

		Inhibitor Pattern Observed[a]			
		For Varied [AX]		For Varied [B]	
Mechanism	Product Used As Inhibitor	At Unsaturated [B]	At Saturated [B]	At Unsaturated [AX]	At Saturated [AX]
Compulsory ordered with [AX] binding first	BX	N	U	N	N
Compulsory ordered with [AX] binding first	A	C	C	N	—
Compulsory ordered with [B] binding first	BX	N	—	C	C
Compulsory ordered with [B] binding first	A	N	N	N	U
Random ordered	A	C	—	C	—
Random ordered	BX	C	—	C	—
Double displacement	A	N	—	C	C
Double displacement	BX	C	C	N	—

[a]C, competitive; N, noncompetitive; U, uncompetitive; —, no inhibition.

Rather than rederiving these relationships, we present them as diagnostic tools for determining the mechanism of reaction. The patterns are summarized in Tables 11.2 and 11.3 for dead-end and product inhibition, respectively. By measuring the initial velocity of the reaction in the presence of several concentrations of inhibitor, and varying separately the concentrations of AX and B, one can identify the reaction mechanism from the pattern of doublereciprocal plots and reference to these tables.

11.4 ISOTOPE EXCHANGE STUDIES FOR DISTINGUISHING REACTION MECHANISMS

An alternative means of distinguishing among reaction mechanisms is to look at the rate of exchange between a radiolabeled substrate and a product molecule under equilibrium conditions

The first, and simplest mechanistic test using isotope exchange is to ask whether exchange of label can occur between a substrate and product in the presence of enzyme, but in the absence of the second substrate. Looking over the various reaction schemes presented in this chapter, it became obvious that such an exchange could take place only for a double-displacement reaction:

$$E + A^*X \rightleftharpoons E \cdot A^*X \rightleftharpoons EX \cdot A^* \overset{-A^*}{\rightleftharpoons} EX \overset{B}{\rightleftharpoons} EX \cdot B \rightleftharpoons E \cdot BX \rightleftharpoons E + BX$$

For random or compulsory ordered reactions, the need to proceed through the ternary complex before initial product release would prevent the incorporation of radiolabel into one product in the absence of the second substrate.

Next, let us consider what happens when the rate of isotope exchange is measured under equilibrium conditions for a general group transfer reaction:

$$AX + B \rightleftharpoons A + BX$$

Under these conditions the forward and reverse reaction rates are equivalent, and the equilibrium constant is given by:

$$K_{eq} = \frac{[BX][A]}{[AX][B]} \tag{11.10}$$

If under these conditions radiolabeled substrate B is introduced in an amount so small that it is insufficient to significantly perturb the equilibrium, the rate of formation of labeled BX can be measured. The measurement is repeated at increasing concentrations of A and AX, to keep the ratio [A]/[AX] constant (i.e., to avoid a shift in the position of the equilibrium). As the amounts of A and AX are changed, the rate of radiolabel incorporation into BX will be affected.

Suppose that the reaction proceeds through a compulsory ordered mechanism in which B is the first substrate to bind to the enzyme and BX is the last product to be released. If this is the case, the rate of radiolabel incorporation into BX will initially increase as the concentrations of A and AX are increased.

As the concentrations of A and AX increase further, however, the formation of the ternary complexes E ·AX·B and E ·A·BX will be favored, while dissociation of the EB and EBX complexes will be disfavored. This will have the effect of the EAX and EA forms. The EAX form will react with B, leading to formation of BX, while the EA form will not. Hence, the rate of radiolabel incorporation into BX will increase with increasing [AX] as a hyperbolic function (Figure 11.5B). The same hyperbolic relationship would also be observed for a reaction that proceeded through a random ordered mechanism. In this latter case, however, the hyperbolic relationship also would be seen for experiments performed with labeled AX and varying [B].

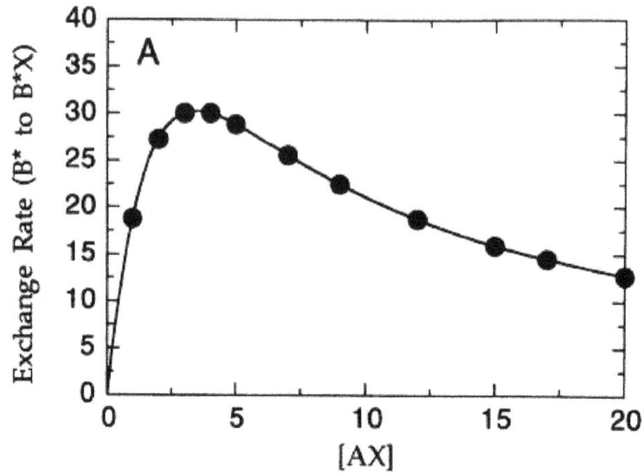

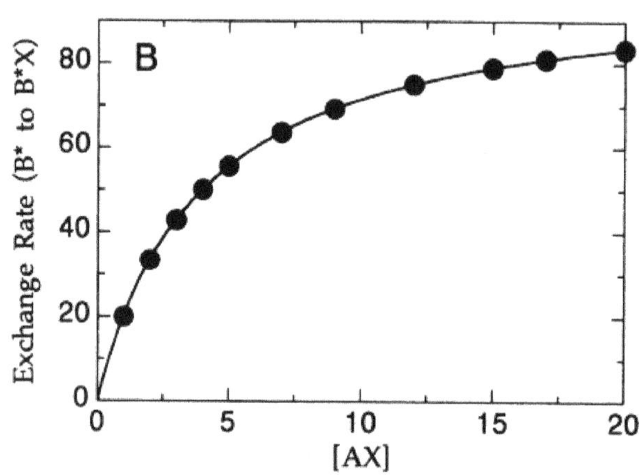

Figure 11.5 Plots of the equilibrium rate of radioisotope exchange between B and BX as a function of [AX] for (A) a compulsory ordered bi bi reaction in which B is the first substrate to bind to the enzyme and BX is the last product to be released, and (B) either a compulsory ordered bi bi reaction in which AX binds first or a random ordered bi bi reaction.

Thus isotope exchange in the absence of the second substrate is diagnostic of a double-displacement reaction, while compulsory ordered and random ordered reactions can be distinguished on the basis of the relation of the rate of radiolabel exchange between one substrate and product of the reaction to the concentration of the other substrate and product under equilibrium conditions.

Chapter 12

FUTURE OF ENZYMOLOGY

There can be no doubt that study of enzymes will continue to occupy the prime position in modern biology (and chemistry!) . This much is amply obvious from each and every context and examples that we have come across in the preceding chapters. The study of enzymes in isolation, most often in purified form, has occupied much of the time in this field.

However, as we have seen in the last chapter, the importance of understanding enzyme function in vivo is very much appreciated now; this will form one of the frontiers in enzymology. Present emphasis on systems biology is a pointer in this direction. Enzymes in sequence, in combination with other enzymes and other cellular components, bring in interesting features often not manifested by an enzyme in isolation – coupled reactions, regulatory networks, and distributed control of metabolism are some of them.

Major metabolic pathways and the enzymes that function in them are well documented. But many more novel reactions and corresponding enzyme catalysts are being discovered on an almost regular basis. For instance, let us consider a few recent representative cases:

1. Most bacteria and all archaea synthesize glutaminyl-tRNA indirectly. The glutamate charged onto tRNAGln is converted to glutamine in the second step by a tRNAdependent amidotransferase. The two enzymes formthe "glutamine transamidosome" that also involves channeling of ammonia .
2. The dehydroalanine moieties of the lantibiotic nisin are formed from the Ser residues of the peptide by the action of lantibiotic dehydratase. This enzyme glutamylates Ser side chains using glutamyl-tRNAGlu for activation;

subsequent glutamate elimination results in the dehydroalanine formation .

3. The biosynthesis of lincomycin A (a sulfur-containing lincosamide antibiotic) recruits two bacterial thiols (mycothiol and ergothioneine) where mycothiol acts as the sulfur donor after thiol exchange .
4. Riboflavin (vitamin B2) is a well-known redox cofactor in a wide variety of flavoproteins. Addition of a fourth ring to its existing three-ring system generates a riboflavin derivative . This previously unknown cofactor catalyzes new types of chemistry and is crucial for the decarboxylation of an intermediate in coenzyme Q biosynthesis.
5. A SAM-dependent enzyme-catalyzed pericyclic transformation leads to the formation of the natural product leporin. Such novel roles for SAM (S-adenosyl-Lmethionine) are likely to be found in other examples of enzyme catalysis .

These are just a few examples that represent author's personal bias. But certainly, many more novel reactions and enzymes will continue to be discovered and reported. We will now consider aspects of enzymology where rapid progress is being made. Understanding the rate enhancements of enzymes continues to be a fundamental challenge of mechanistic enzymology . These and other topics are anticipated to attract much attention of enzymologists in the foreseeable future.

12.1 TRANSITION-STATE ANALYSIS AND COMPUTATIONAL ENZYMOLOGY

Conventional view of enzyme catalysis treats the transition state (TS) in thermodynamic terms. Equilibrium between reactants and the TS is assumed here; tight binding to the enzyme active site would sequester the TS from solution and increase the reaction rate. Kinetic isotope effects can provide direct information on the enzymatic TS . When the intrinsic kinetic isotope effects (KIEs) are available for an enzyme reaction, then the TS structure can be deduced in the usual physical organic chemistry sense. Interpretations of KIEs give detailed bond order and geometric features of the transition state for an enzyme.

In turn, molecular electrostatic potential surfaces of these transition-state depictions guide chemical synthesis of transition-state analogs, yielding excellent, high-affinity inhibitors . Complexes of such inhibitors with their corresponding enzymes provide structural models for computational analysis of enzymatic transition states. Taken together, KIEs and computational enzymology provide a conceptually complete picture of the TS in an enzymatic reaction.

There is a progressive appreciation that the enzymatic transition state is a state of maximum free energy. And this may not be simply captured as an equilibrium between the Michaelis complex, transition state, and products. Dynamic contributions of protein motion must be incorporated to provide a full understanding of enzyme catalysis. Besides applications of kinetic measurements, isotope effects, time-resolved spectroscopy, NMR, and X-ray crystallography, computational enzymology is making significant contributions in understanding the enzyme TS. Most experimental techniques have focused on conformational changes that occur in microsecond to millisecond timescales; these times are often correlated with catalysis.

These timescales are much slower than the TS lifetimes that occur on a bondvibrational timescale. Chemical bond-forming/bond-breaking steps of interest (in enzyme catalysis) have vibrational modes in the low femtosecond timescales (Fig. 12.3). They are valuable in understanding TS but otherwise are experimentally difficult to access. This is where developments in computational enzymology are making some headway. Bond vibrations during enzyme catalysis can be simulated with some accuracy in the femtosecond to the nanosecond timescale. This provides computational dynamic access to the timescales of the TS lifetimes. However, commonly encountered enzymatic catalysis timescales (10_2 s) require enormous computational time; and this is far beyond continuous dynamic calculations possible at present. This frontier in computational enzymology and of complex enzyme models was recognized with chemistry Nobel Prize in 2013 . The bemergence of the quantum mechanical/molecular mechanics (QM/MM) approach allows one to ask what the origins of the catalytic power of enzymes actually are.

The enormous increase in computer power makes it virtually certain that computer simulations will increasingly contribute in modeling molecular enzyme catalysis. This approach also promises to capture the contributions of subtle protein molecular dynamics to enzyme rate accelerations.

12.2 SINGLE-MOLECULE ENZYMOLOGY

The Michaelis-Menten equation is a highly satisfactory description of kinetic data involving a very large number of enzyme molecules in the assay. However, all enzyme molecules are not synchronized with each other in an ensembleaveraged kinetic measurement. The extraction of dynamic information from such an asynchronous assembly is complicated and difficult. A single enzyme molecule gives kinetic signals that reflect the dynamic states of that individual catalytic entity. This dynamic information is lost in the average signals of the ensembles. Reactions involving single enzyme molecules can now be examined by the advances in fluorescence and other time-resolved spectroscopic techniques . These techniques combined with computational approaches allow real-time access to study the dynamic behaviors of individual molecules.

Would the ensemble-averaged enzyme kinetics also hold for a single enzyme molecule? This interesting question has attracted serious attention in the recent times for the following reasons. Single-molecule behavior is a particularly powerful way of uncovering (a) mechanistic pathways and intermediates, (b) how enzyme conformational fluctuations affect catalytic activity, and (c) heterogeneities hidden in the ensemble average. The turnover events of a single enzyme molecule are intrinsically stochastic. We find the sequence of the time intervals between consecutive turnover events (the waiting times, denoted as τ), as a function of the turnover index number.

The average waiting time (denoted by hτi) of a single enzyme molecule plotted against 1/[S] mimics the linear Lineweaver–Burk plot (recorded in an ensemble measurement. The rate equation for single enzyme molecule kinetics is shown below.

$$\frac{1}{\langle\tau\rangle} = \frac{\chi^2[S]}{C_M + [S]} \quad \left(\text{compare this with } v = \frac{k_{cat}[E_t][S]}{K_M + [S]}\right)$$

The reciprocal mean waiting time determined for a single enzyme molecule is thus related to enzyme-catalyzed velocity in an ensemble measurement (1/hτi ¼ v/[Et]). While the Michaelis–Menten relation continues to hold for the single enzyme molecule kinetics, interpretations of kinetic constants are different. Specifically, the kcat (formally represented as χ2 for single enzyme study) measured at saturating [S] is a weighted harmonic mean of the different catalytic turnover rate constants represented in the single enzyme over time. Similarly, KM (¼ (kcat + k_1)/k1 or CM for a single molecule) also acquires an ensemble-averaged meaning .

We notice that single-molecule and ensemble-averaged Michaelis–Menten kinetics can be reconciled . An interesting insight fromsinglemolecule study is that the waiting times measured at high [S] levels follow an asymmetric probability distribution. The long time-span of the catalytic turnover rate constant (i.e., kcat) indicates that the single enzyme molecule's catalytic velocity fluctuates over a broad range (from 10^{-3} to 10 s) of timescales. These catalytic fluctuations point to conformational isomers of an individual enzyme slowly interconverting over time.

Conformational heterogeneity has been experimentally observed in the studies of cholesterol oxidase, staphylococcal nuclease, and a few other enzymes. A single enzyme molecule displays inevitable, stochastic fluctuations in its catalytic activity. The effects of such fluctuations would be less significant for a system comprising of large number of enzyme molecules. However, many processes inside cells rely on the activity of a single enzyme molecule, such as in DNA replication, transcription, translation, and protein transport along the cytoskeleton. Stochastic fluctuations due to low copy number of enzymes have important physiological implications for cells/ organelles. These are now being probed on a single-molecule basis in vivo.

12.3 STRUCTURE-FUNCTION DISSECTION OF ENZYME CATALYSIS

As discussed earlier in this book , both conformational flexibility and protein motion are very important for enzyme catalysis. A structure–function approach to understand enzyme action uses a combination of kinetic (including rapid, transient kinetics) and structural techniques. The ultimate goal of such a study is to develop an in-depth mechanistic understanding of enzyme function. The field has traditionally made use of protein chemical modifications, spectroscopic tools (like fluorescence, circular dichroism, etc.), and more recently molecular biology tools (like generation of site-directed mutants, truncated or chimeric enzyme proteins). Whereas solution NMR has provided some information on enzyme molecular dynamics (although at timescales much slower than most events in enzyme catalysis), as of now this structural tool is limited to small-sized proteins.

Mass spectrometry of larger polypeptides and oligomeric proteins is also coming of age. Much of direct structural information on enzymes has come from X-ray crystallography – mostly presenting to us a static picture of the enzyme. As aptly stated by the late Jeremy Knowles (biochemist and professor at Harvard University) – "studying the photograph of a racehorse cannot tell you how fast it can run"; and thus there is a limit to what a snapshot protein structure can reveal. Although snapshots of enzyme bound to substrate, product, or transition-state analogs are valuable, they do not capture structural dynamics of catalytic action. Nevertheless, presence of molecular tunnels in enzymes like tryptophan synthase would not have been apparent without detailed structural inputs.

Overwhelming biochemical, mechanistic, and mutational data for any enzyme study must be supplemented by structural approaches. More recent developments in computational structure prediction (see below) provide complementary inputs. Structure-based drug discovery has also benefited from these approaches. One remarkable success in this venture is the development of HIV protease inhibitors . These inhibitors were rationally designed from the knowledge of the structure and mode of action of aspartyl proteases. The discovery of saquinavir, the first protease inhibitor, made use of the promising transition-state mimic chemistry.

A fine balance between structural rigidity and conformational plasticity results in the unique catalytic power of enzymes. And structural enzymology also aims to address catalytic motions in detail . How enzymes achieve a catalytically competent state has become approachable only recently through experiments and computation.

Site-Directed Mutagenesis and Crystal Structures Application of molecular biology tools to probe enzyme function has matured over the last couple of decades. The reader may refer to standard textbooks of molecular biology and many protocol/ recipe books on how to construct site-directed mutants and to genetically engineer enzymes. The relevant cDNA and an expression system to obtain the mutant enzyme are all that are required . The native and various mutant enzyme forms are then subjected to rigorous structural analysis through X-ray crystallography, circular dichroism spectra, and other tools. Excellent insights continue to be gathered on residues critical for catalysis, binding, and structural stability/flexibility for many enzymes.

Site-directed mutagenesis offers a powerful approach to rationally modify an enzyme. It enables enzymologists to selectively replace active site residue (or any others) and ask some very interesting mechanistic questions. Yet site-directed mutagenesis does not fully account for enzymatic catalysis, because the effects of individual substitutions on catalysis are neither additive nor independent. At the resolution of amino acid residue level, one can check whether a given residue is relevant for binding and/or catalysis. This may not give unambiguous answers as, for many enzymes, substrate binding and catalytic residues often overlap; and they may not be clearly demarcated in the enzyme active site. In fact, few examples show that specificity can reside beyond the amino acids directly interacting with the substrate – requiring major structural changes with loop grafting, etc. Nature presents us with 20 naturally occurring amino acids in proteins; an opportunity exists to replace a given residue by any one of the other 19. But still the choice is limited because very few of these substitutions are more conservative than others. For instance, Val, Ile, and Leu are often interchangeably accepted; also, a Glu residue may replace Asp and vice versa. Alanine scanning mutagenesis is a mature tool and is often used to determine the contribution of a specific residue to the stability and/or function of a given protein. Alanine is chosen because its R group (methyl) is least disruptive and imitates the secondary structure preferences of many other amino acids. Such Ala replacements are usually done by site-directed mutagenesis or generated randomly by creating a PCR library.

Standard site-directed mutagenesis is largely limited to 20 natural, proteinogenic amino acid residues. Nonnatural side chains often provide mechanistic insights. One approach has been to replace the relevant residue by Cys (through SDM) and then alkylate that Cys-SH by a suitable reagent. Interesting structural variations of imidazole side chains, with subtle pKa changes, could then be tested . In another example, the catalytic activity of glutamine synthetase R!C mutant was rescued back by chemically modifying the Cys-SH back to an arginine analog by covalent modification with 2-chloroacetamidine . More recently, sophisticated tools and technology are in place to directly incorporate nonprotein amino acids into proteins in a position-specific manner. Nonnatural amino acids may be introduced into proteins by manipulating in vitro protein translation as well as through in vivo strategies by expanding the genetic code . Pyrrolysine, the twenty-second protein amino acid,was found at the active site of methyltransferases from methane-producing archaea. Like the 20 common amino acids, pyrrolysine is synthesized in the cytoplasm and incorporated at a specific position during the translation of the growing polypeptide chai. The pyrrolysine biosynthetic cassette could be used to incorporate other useful modified amino-acid residues into proteins.

Examples of SDM to probe enzyme function have rapidly grown. We will look at only a few case studies to illustrate main issues and highlight some difficulties with this approach. The first significant effort to change the substrate specificity of an enzyme was with L-lactate dehydrogenase.

- ☆ Holbrook's group achieved a highly active, malate-specific dehydrogenase by redesigning Bacillus stearothermophilus lactate dehydrogenase framework . This involved three amino acid replacements, namely, D197N, T246G, and Q102R.
- ☆ Human arginase II is highly specific and acts on arginine to produce ornithine and urea. A single amino acid replacement (the N149D variant) converts this enzyme into an agmatinase with almost no activity on arginine .
- ☆ Glucose isomerase site-directed mutagenesis has resulted in enzyme forms with altered pH optima and altered divalent metal ion specificity.
- ☆ The rational modification of enzymes to change or extend their coenzyme specificity has also been possible . The NADP-specific glutathione reductase was altered into a NAD-dependent enzyme. Similar efforts in other dehydrogenases such as glutamate dehydrogenase are reported.
- ☆ There are many attempts to rationalize the enzyme stability as a useful parameter through site-directed mutagenesis . These have been largely empirical because we do not yet fully understand what contributes to protein stability.
- ☆ Site-directed mutagenesis provides enzymologists an opportunity to tinker with active site residues and ask some really interesting mechanistic questions. Almost every position in the subtilisin sequence has been subjected to SDM. The three critical active site catalytic residues S221, H64, and D32 of subtilisin were evaluated through this approach. For the subtilisin S221A mutant (which still retains some activity!), the reaction cannot proceed by the usual serine acyl-enzyme intermediate (ping-pong mechanism). While

such a mechanistic change from an acyl-enzyme to a direct water attack may be possible, the converse is much more difficult because the essential nucleophile would be missing. It is very easy to lose enzyme function but much more difficult to gain a new function through SDM!

- ☆ The H95 residue of triosephosphate isomerase is important for catalyzing the enolization of the substrates. The H95Q mutant is impaired in its ability to stabilize this reaction intermediate. There is an associated change in the proton transfer pathways mediated by the mutant enzyme .
- ☆ The cancer-associated isocitrate dehydrogenase mutant (R132H form of IDH1 isozyme) loses its native function but is able to catalyze the NADPH-dependent reduction of 2-oxoglutarate to R(-)-2-hydroxyglutarate (D-2-hydroxyglutarate) .

Site-specific mutagenesis approach has undoubtedly extended our knowledge of enzyme mechanism and function. The necessary first step in the kinetic analysis of a mutant enzyme is to show that any observed change in catalytic activity is solely because of the targeted alteration. For this, the enzymologist must be prepared to analyze both the structural and functional consequences of the mutation(s) made. To begin with, this should include a thorough evaluation of kinetic parameters like KM, Vmax, kcat, and kcat/KM. Gross structural changes in a mutant enzyme can be discerned through techniques like circular dichroism and gel filtration chromatography.

X-ray crystallography of the native and mutant enzyme forms (as well as their frozen structures bound to substrate, product, inhibitors, and transition-state analogs) offers valuable information. Of course, the pretty structures should conform to the hard data . It is a distinct possibility that mutant enzymes might follow a different reaction pathway. This was highlighted above, with examples of triosephosphate isomerase and serine proteases. Therefore, a detailed mechanistic analysis is routinely needed for proper appreciation of the effects from site-directed mutations.

Changing Landscape of Enzyme Allostery Historically, understanding the control of enzyme activity through feedback mechanisms led to the simultaneous discovery of enzyme cooperativity and allostery. Feedback inhibition and cooperativity over the years have got connected and appear to be two faces of the same coin. We have seen their implications to metabolic regulation in the previous chapter . The non-Michaelian kinetic feature allows an enzyme to function as a concentration-dependent metabolic switch. Classical models invoke distinct enzyme conformations associated with allostery.

These may involve multiple ligand binding sites (with associated conformational selection and induced fit) or polypeptide oligomerization. In most allosteric enzymes, the allosteric binding site for the ligand lies far away from the active site. Therefore, structural communication paths (via the levers and pulleys of protein structure) must exist between these sites. First mechanistic description for allosteric regulation was proposed more than sixty years ago; other possible mechanisms are being advanced regularly to describe these phenomena . Investigations at atomic detail (by high-resolution NMR of specifically labeled side chains) of glucokinase (a monomeric enzyme) for its kinetic cooperativity have provided recent insights. The enzyme molecule samples a number of conformational states in the absence of glucose. However, this population of conformations shifts toward a narrow, well-structured ensemble of states in the presence of glucose .

Allostery is not always mediated by conformational changes that can be detected by standard techniques like X-ray and NMR relaxation measurements, determination of H/D exchange rates, and isothermal titration calorimetry (ITC) experiments. These methods give a time-averaged snapshot of the protein 3D structure. However, recent advances in spectroscopy (probes to explore time-resolved dynamics of protein conformational changes) and the computational approaches (to study molecular dynamics simulations) indicate that multiple conformational states of the enzyme exist even in kinetically simple Michaelis complexes. Also, the free enzyme itself is a collage of protein conformational states. This protein disorder is clearly observed in the dynamic motions as measured by distance- and time-resolved NMR studies. Present-day structural biology offers unequivocal evidence of multiple conformations in preexisting equilibrium for glucokinase, trypsin-like proteases, maltose-binding protein, etc. The new outlook on allostery incorporates this more dynamic view of the enzyme protein. Accordingly, allosteric control may manifest by a population shift in the statistical ensembles of many states, with some regions of low local stability and others of high stability. Ligand binding affects the relative free energies of these states; they in turn differ in their affinities for other ligands and/or their activity. This is different from the earlier concept of a few well-defined static conformational states (such as R and T states). Various mechanisms of allostery described to date are summarized in the Table 12.1.

The new view of allostery encompasses a conformationally dynamic continuum of allosteric phenomena. With recent discoveries, we have moved towards increasing enzyme protein dynamics (disorder or fluctuations) starting from (a) rigid body movements to (b) side-chain dynamics, (c) backbone dynamics, (d) local unfolding, and finally (e) intrinsically disordered structures.

Table 12.1 Molecular mechanisms of allosteric regulation

Mechanism	Examples
Closure/opening of active site	3-Phosphoglycerate dehydrogenase (the active site cleft closes upon binding of the end product, Ser)
Changes in active site conformation	3-Deoxy-D-arabinoheptulosonate-7-phosphate synthase (aromatic amino acid (the end product) binding leads to minor conformational modifications and prevents substrate binding)
Change electrostatic properties of active site	Chorismate mutase (aromatic amino acid (the end product) binding brings a Glu residue into the active site; causes a major change in its electrostatics and repels the negatively charged substrate)
Influence protein–protein complex formation	ATP phosphoribosyltransferase (His (the end product) binding converts an active dimer to an inactive hexamer)
Affect protein flexibility	Dihydrodipicolinate synthase (Lys (the allosteric ligand) binding affects distant sites via a change in the protein vibrational modes)
Shift in ensemble of conformer population	–Glucokinase (its intrinsically disordered small domain samples a broad conformational ensemble; upon glucose binding, the population shifts toward a narrow, well-ordered ensemble)
	–Dimeric catabolite activator protein (binding of the first cAMP molecule lowers the affinity for the second cAMP molecule; first cAMP binding enhances motions within the protein, whereas the binding of second cAMP decreases these motions and flexibility

Allostery, by definition, involves the propagation of signals between different sites in a protein. This may occur in the absence of detectable conformational changes and may be exclusively mediated by transmitted changes in protein motions. The fact that the change in dynamics occurs in the absence of significant structural change suggests that dynamics alone may convey allosteric information.

There is much interest to probe the existence of an entire channel/network of amino acids through which allosteric signals are communicated. One approach that could map such paths (perturbations that travel across the structure) and implicate the interacting amino acid residues is through measurements of a double-mutant cycle.

Energetics of such residue interactions allows us to infer the degree of functional coupling between different sites of a protein. A complementary approach uses a sequence-based statistical analysis (Statistical Coupling Analysis, SCA) method for elucidating the architecture of functional couplings in proteins . If two residues in a protein are functionally coupled, then they should have coevolved. This coevolution can be scored by statistically comparing homologous protein sequences. A combination of SCA and double-mutant cycles along with functional, structural, and folding analyses can provide insights into the existence of an entire wave, wire, channel, or network of amino acid residues through which allosteric signals are transmitted.

Transplantation of allosteric regulation has been possible in hemoglobins and glycerol kinase . Very few residues may be involved in the manifestation of allosteric regulation. The allosteric features of crocodile hemoglobin could be introduced into human hemoglobin by substitutions at 12 amino acid residues. Changing only 11 of 501 total residues (about 2%) converts an unregulated glycerol kinase to an allosterically regulated glycerol kinase. Small molecules can exert strong effects from unexpected locations, and hence searching for new allosteric sites in enzymes is a challenge. It is hard enough to predict what ligand might bind in a binding pocket, but the presence of allosteric binding sites further complicates the matter. It may become possible to develop algorithms to distinguish allosteric from active sites with better databases in the future. It may also be feasible to predict allosteric regulation from protein structural data .

The emerging radical view of enzyme function is that each catalytic step corresponds to an ensemble of thermodynamic and structural states. Incidentally, allostery and catalysis no longer appear as distinct phenomena but as the manifestations of the same intrinsic protein dynamics. Important issues for further research in allosteric enzymes include (a) the mechanisms by which an allosteric effect is transmitted via amino acid networks, (b) how the distribution of protein conformations is altered, and (c) the timescales at which the redistribution of these conformations occurs.

Predicting Enzyme Structure and Function Computational tools have revolutionized the whole of biology. It is no wonder that substantial progress has taken place in enzymology, one of the more quantitative of the biological sciences.

Incorporating new tools and technology to understand enzyme catalysis is a recurring theme in enzymology. Computational enzymology has joined the earlier methods involving kinetic measurements, kinetic isotope effects, crystallography, and distance-/time-resolved NMR. We have noted above that computational enzymology has made significant inroads into transition-state analysis and molecular dynamics of enzyme action. Besides these, there is a rapid move toward enzyme structure and function prediction through computational approaches. This need has arisen as we (a) accumulate a large number of sequenced genomes; (b) come across orphan open reading frames, with no clues of their function; and (c) express sequences into proteins and even crystallize them without their actual functional demonstration .

Assigning valid functions to unknown (putative!) proteins/enzymes identified in genome projects is a challenge . While experimental testing remains essential, computational approaches can help guide this experimental design. Bioinformatics approaches are being perfected to (a) identify informative sequence relationships using structure and genome context, (b) allow accurate highthroughput structure prediction through homology modeling, and (c) dock metabolites in silico to provide accurate and testable list of potential enzymes.

Microbial metabolic pathways often are encoded by genome neighborhoods (synteny and associated gene clusters and/or operons). Such positional information can provide important clues for enzyme function assignment. For instance, pathway docking is an efficient strategy for predicting in vitro enzyme activities and allocating in vivo physiological functions. It has been possible to identify novel metabolites, enzyme activities, and biochemical pathways through this tactic.

In view of the exponential growth in genome sequence data sets (with significant proportion of sequences with unknown enzyme functions), an integrated strategy for functional assignment was recently proposed . This enzyme function initiative (EFI; the multicentric program under the National Institute of General Medical Sciences, USA) looks to predict the substrate specificities of unknown members of mechanistically diverse enzyme superfamilies – thereby predicting their functions. The approach exploits conserved features within a given superfamily such as known chemistry, identity of active site functional groups, and composition of specificity-determining residues/motifs/structures. Initial enzyme targets chosen for this purpose include members of the amidohydrolase, enolase, glutathione transferase, haloalkanoic acid dehalogenase, and isoprenoid synthase superfamilies. Members of these enzyme superfamilies are functionally diverse (conserved partial reactions or chemical capability but with divergent overall function) which makes functional assignment difficult. Homology inferred from simple sequence comparisons alone cannot guide functional assignment in such situations. Therefore, an integrated approach involving the following components is proposed:

- Perform bioinformatic analysis to cluster sequences into probable isofunctional groups; assign tentative functions for further investigation by structure determination, structural modeling/docking, and biochemical experimentation.
- Carry out homology modeling to expand the use of structural models; thereby guide functional assignment to proteins without experimentally determined structures.
- Employ computational docking methods to leverage structure and guide functional assignment by suggesting substrates/ligands for biochemical experimentation.

12.4 DESIGNING NOVEL CATALYSTS

Curiosity and the desire to imitate general principles of biological catalysis have led to many developments in the design and construction of artificial enzymes. Present approaches to create novel catalysts fall into three general categories – (a) de novo design and synthesis of catalysts from polypeptides and nonprotein building blocks (such as macrocyclic compounds, poly-ethyleneimine, synthetic chemical/genetic polymers, and cyclodextrins); (b) modification/evolution of existing catalysts such as protein enzymes or ribozymes, by genetic or chemical methods; and (c) designer enzymes made to order. While antibodies and RNA as catalysts will be discussed subsequently, we will first focus on chemical models, enzyme mimics, and hybrid catalysts.

Chemical Models, Enzyme Mimics, and Hybrid Catalysts Enzyme models of increasing complexity have been designed and discovered. A small chiral moleculelike proline may be considered the simplest enzyme. Recent reports on asymmetric catalysis by proline and its derivatives include activation of carbonyl compounds via nucleophilic enamine intermediate. Several highly enantioselective important carbon–carbon bond-forming reactions (aldol additions and Mannich reaction) have been developed using this approach. Many model systems anticipated their enzyme counterparts much ahead of time. Otto Warburg studied the oxidation of unsaturated fatty acids by combined action of iron and sulfhydryl (-SH) groups in 1925. Lipoxygenases, containing iron and –SH groups essential for their oxidative activity, were discovered much later. Aniline-catalyzed rapid decarboxylation of acetoacetate via the "aniline–acetoacetate complex" is exemplified by JBS Haldane in his 1930 book on enzymes. More recently, chemical hydrogenase mimics with Co and Ni centers are reported for exploiting hydrogen as fuel. Models for enzymes performing free radical chemistry and other redox reactions are being sought to be incorporated in clean energy programs. The study of glutathione peroxidase anticipated the small molecular enzyme model ebselen as well as an antibody enzyme – abzyme .

As pointed out before, there are two reasons to study enzyme constructs and mimics. One objective of chemists and biologists is to elucidate molecular basis of enzyme function. Secondly, using the available knowledge base (which is still far from complete!), one could attempt to design and build novel catalysts – the so-called tailor-made enzymes.

The chemical alteration of an existing enzyme by introducing additional functional groups is one route to rational enzyme design. The semisynthetic enzyme so generated can display very different catalytic activities from that of the parent enzyme. For example, papain (a thiol protease) was converted to an effective redox catalyst (an oxidoreductase) by appropriately attaching a flavin to the enzyme sulfhydryl group. In another example, new binding domains are selectively introduced to build/alter enzyme specificity. RNase A is a relatively nonselective enzyme hydrolyzing phosphodiester linkages of RNA. It was made specific to a definite RNA sequence by creating a chimeric RNase A – wherein a covalently attached single-stranded DNA confers specificity by annealing at the complementary RNA sequence (Fig. 12.1).

The oligonucleotide-tagged RNase A above is an effort in modifying the substrate specificity of an enzyme. However, another attempt was made to build a restriction endonuclease from first principles. Several chemical agents intrinsically possess DNA cleavage activity. If these are incorporated into DNA/RNA binding proteins, a specific nuclease may be created. Indeed, one could design synthetic hybrid molecules with two components – one for specific DNA sequence recognition and the other to cleave the DNA adjacent to it. The technique called affinity cleavage is based on the construction of such bifunctional molecules (Fig. 12.2). Many artificial metalloenzymes have emerged from similar approaches. Chemistry of EDTA-Fe[II] with oxygen is an excellent source of hydroxyl (HO•) radicals for DNA cleavage.

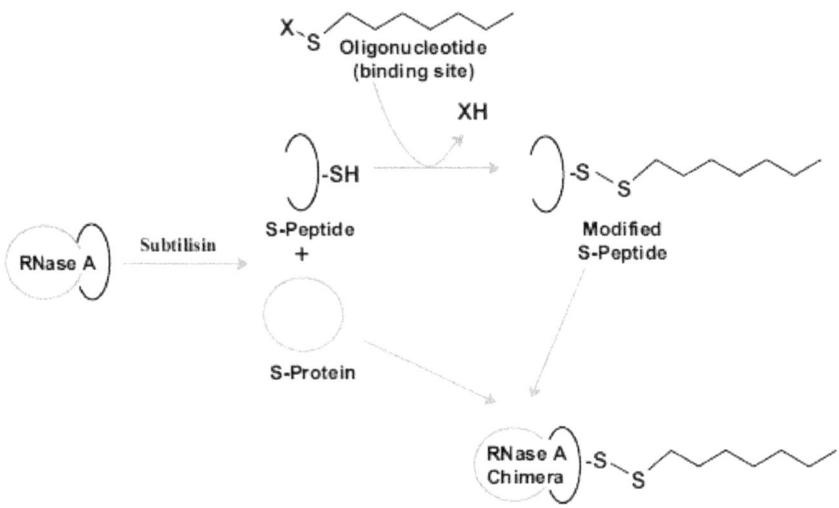

Fig. 12.1 Construction of a sequence-specific RNase A chimera. The S-peptide was modified with an oligonucleotide sequence. The functional RNase A was reconstituted by combining the modified S-peptide with the S-protein.

This reaction can be turned into a catalytic cycle by adding ascorbate so that Fe[III] formed is reduced back to Fe[II]. Several chemical agents possess nuclease activity; rare earth metal ions are active in hydrolyzing phosphodiester linkages in DNA. For instance, DNA scission by Ce[IV] is hydrolytic and not oxidative. The fragments so generated can be religated using a ligase. The sequence-recognizing moiety brings the molecular DNA scissors close to the target phosphodiester bond. Binding to specific DNA sequences is provided either by a short oligonucleotide or by a specific DNA-binding domain of the protein. Sequence recognition has also been achieved by pcPNA (a pseudo-complementary peptide nucleic acid).

Most restriction enzymes used in molecular biology recognize a stretch of six bases and cut DNA at that site. Assuming random distribution of A, G, T, and Cs, there is a high probability (46, which is once every 4096 bp of DNA) of finding these restriction sites in DNA. To cut at a single unique position in human genome, we could use a 16mer (or longer) sequence-recognizing moiety (416; for >109 bp of DNA in human genome!). This site selectivity compares well with the molecular mtools like (a) the Achilles' heel cleavage (only one of the many restriction sites in a genome is specifically protected from inactivation by a cognate methyltransferase; this in turn creates a unique restriction enzyme cleavage site) and (b) the genome editing by the CRISPR/Cas system. An artificial restriction enzyme that can cut only at one position in the human genome will require a 16mer (or longer) sequencerecognizing moiety.

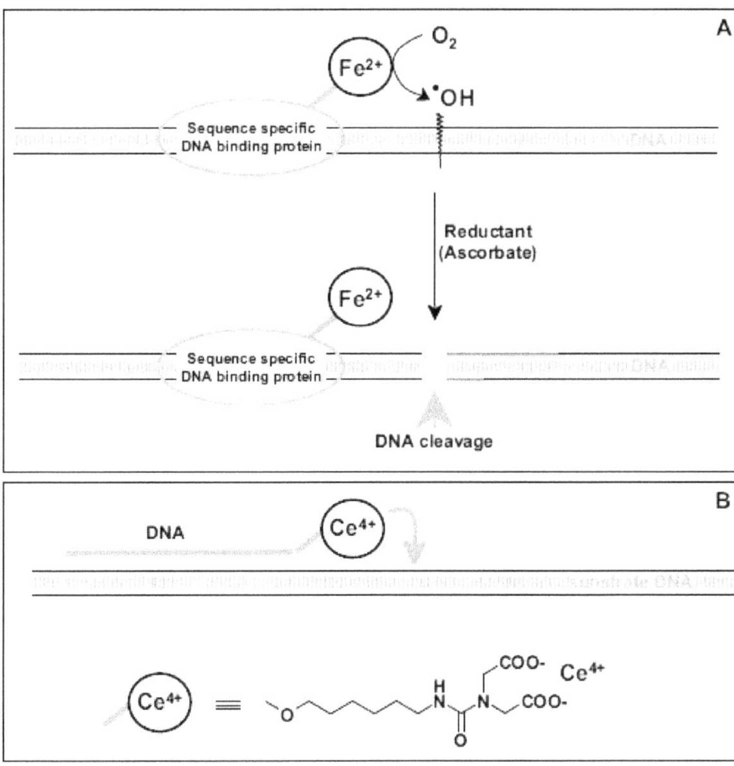

Fig. 12.2 Construction of artificial restriction enzymes (A) Fenton chemistry of EDTA-Fe[II] with oxygen generates hydroxyl (HO•) radicals for local DNA cleavage. Specificity is provided by using a unique sequence-recognizing moiety (e.g., trp or lac repressor or catabolite activator protein of E. coli). (B) A restriction enzyme construct of Ce4+ complex of iminodiacetate. This Ce4+ nuclease is juxtaposed to the specified DNA sequence for cleavage by an oligonucleotide sequencerecognizing moiety

Novel catalysts may also be crafted on nonprotein molecular framework – including synthetic macromolecules of nonbiological origin . Poly(ethyleneimine) polymers possess intrinsic acid–base groups; along with pyridoxal phosphate or other suitable cofactors, some degree of catalysis was demonstrated.

Breslow's group has made extensive efforts to emulate enzyme catalysis using cyclodextrin scaffolds . While the cyclodextrin cavity offers a binding pocket, additional functional groups attached to the cyclodextrin ring offer new enzyme mimics (synzymes). The cavity size can be varied by choosing either β-cyclodextrin (seven glucose units in the ring) or γ-cyclodextrin (eight glucose units in the ring). Mimics of RNase A (β-cyclodextrin bis-imidazole) and chymotrypsin (for its esterase activity!) are demonstrated. A flexible capped cyclodextrin with the well-fitting substrate afforded a rate acceleration of 8 - 107 fold in p-nitrophenyl ester hydrolysis (Fig. 12.3).

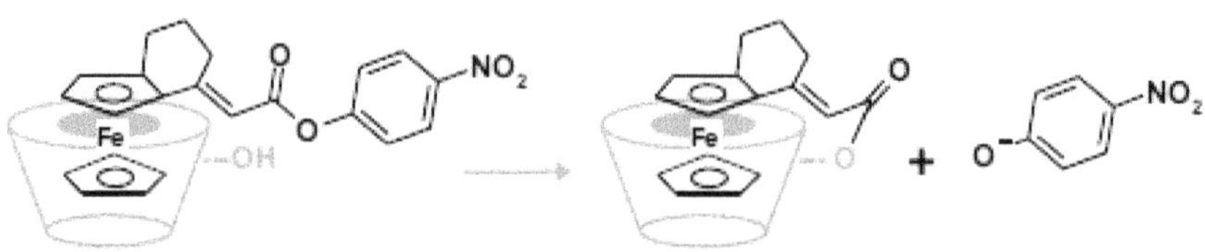

Fig. 12.3 A β-cyclodextrin mimic of chymotrypsin displaying catalysis of p-nitrophenyl ester hydrolysis. The cyclodextrin cavity binds the hydrophobic ferrocene core of the ester substrate

Lastly, a word of caution on rate acceleration observed for enzyme mimics. Most protease/esterase models use p-nitrophenyl esters as substrates. The p-nitrophenyl esters are not protease/esterase substrates "in real life." With p-nitrophenyl esters as model substrates, partly the better leaving group effect contributes to the observed rate accelerations. Leaving groups whose pKas are above 9.0 lead to much lower acceleration. One should account for such a p-nitrophenyl ester syndrome when evaluating the performance of synzymes .

Antibody Catalysts (Abzymes) Linus Pauling recognized in the 1940s that the ability of an enzyme to speed up a chemical reaction arises from the "complementarity of its active site structure to the activated complex (i.e., the transition state)." This has given rise to the productive field of catalytic antibodies (or abzymes).

Antibody molecules represent a class of proteins with high affinity and exquisite selectivity; they could be raised against any small molecule (hapten) of our choice. Antibodies that can bind to transition-state analogs of a substrate should therefore catalyze the conversion of that substrate, through the transition state, to the product. Abzymes production involves the following steps:

- ☆ Generating a stable transition-state analog using molecular design and chemical synthesis
- ☆ Raising antibodies (monoclonal antibodies to be more precise) with the TS analog as the hapten
- ☆ Isolating antibodies which bind to the TS analog as potential catalysts for that reaction

The seminal prediction of Pauling was verified by Lerner's research group in the 1980s. During an ester hydrolysis, the sp2 hybridized carbonyl carbon is converted to a sp3 hybridized carbon in the intermediate; and the carbonyl oxygen resembles an oxyanion. The transition state presumably looks more like this unstable intermediate (sp3, oxyanion). A phosphonate ester mimic, with a sp3 hybridized phosphorous replacing the sp2 hybridized carbonyl carbon (Fig. 12.4), was synthesized. While very resistant to hydrolysis, this phosphonate also has negatively charged oxygen similar to the intermediate during ester hydrolysis. The mouse antibodies (the monoclonal – 6D4) against this phosphonate structure catalyzed the corresponding carboxylic acid ester hydrolysis.

More than 100 interesting examples of abzyme catalysis are known. They include many reactions that cannot be achieved by standard chemical methods . Besides the ester hydrolysis mentioned above, these include pericyclic processes, group transfer reactions, additions and eliminations, redox reactions, aldol condensations, and a few cofactor-dependent transformations. Some of these are listed in Table 12.2 below.

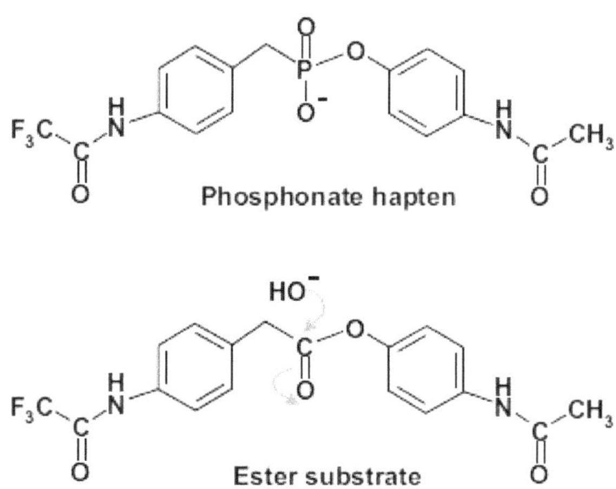

Fig. 12.4 The phosphonate transition-state analog used as hapten. The monoclonal antibodies (abzyme 6D4) against this hapten displayed esterase activity on the corresponding ester shown

Table 12.2 Antibody catalysts generated for different reaction types

Abzyme[a]	Reaction catalyzed
6D4	Carboxylic ester hydrolysis
48G7	*p*-Nitrophenyl ester hydrolysis
1F7	Claisen rearrangement (chorismate to prephenate)
AZ-28	Oxy-Cope rearrangement
39-A11	Diels-Alder reaction
7G12	Ferrochelatase
33F12	Aldolase
34E4	E2 elimination of nitrobenzisoxazole
2F3 (scFv)	Glutathione peroxidase

a. All the abzymes (except 2F3) are monoclonal antibodies . The scFv (single chain Fragment variable) of abzyme 2F3 was activated by chemical modification of a reactive Ser by attaching Se to it

Virtually all experiments with abzymes employ monoclonal antibodies (Table 12.2). The monoclonals are a single homogeneous catalytic species (unlike the polyclonal sera), and their use greatly simplifies kinetic, mechanistic, and structural characterization of an abzyme. While expressing catalytic antibodies (or Fab fragments), folding of the two chains into native state is a challenge. One approach is to express and secrete them into the periplasmic space of E. coli, exploiting its diminished protease activity and ability to correctly fold and form disulfide bonds. The single chain Fragment variable (scFv) version of the antibody exhibits the same catalytic parameters as the parent monoclonal antibody. The scFv is a recombinant protein construct of a VL chain tethered to a VH chain with a polypeptide linker and is expressed efficiently by bacteria.

Fastest enzymes are diffusion limited catalysts. Even the less than perfect ones typically have apparent bimolecular rate constants (kcat/KM) between 106 and 108 $M^{-1} s^{-1}$. Catalytic antibodies have rate accelerations many orders of magnitude below their enzyme rivals. By definition, abzymes are catalytic antibodies specific to (and raised against) the corresponding transition-state mimic. We note that stabilization of the transition state is necessary but not sufficient by itself for achieving good catalysis. Many other factors – like active site functional groups, conformational flexibility, shielding of the reaction intermediates, cofactor needs, etc. – substantially contribute to rate accelerations . Transition states themselves have fleeting lifetimes and cannot be captured. Synthesis of effective TS analogs must therefore draw on our chemical intuition about the conformational, stereochemical, and electronic features of the reaction under study. Since no stable molecule can reproduce all the characteristics of the actual TS, hapten design strategies have focused on incorporating the salient features of the TS. In the sum, limited catalytic ability of abzymes may be attributed to one or more of the following:

- ☆ Antibody scaffold is fixed, and this means limited structural space is explored for catalyst building. Real enzymes come in a variety of structural folds.
- ☆ The polypeptide sequence space exploited during antibody maturation is also limited.
- ☆ Catalytic antibodies may lack structural dynamics necessary for optimal catalysis.
- ☆ TS analogs are imperfect mimics of the actual transition state. The TS mimic designed may not capture the best options for discriminatory binding of the substrate versus TS.

Basic strategy to produce catalytic antibodies is indirect. Here the immune system is directed to evolve not for catalysis but toward binding tightly to an imperfect TS analog. To date, antibody enzymes display only modest catalytic activity and have not found significant practical utility (though a few have reached the market!). Nonetheless they continue to be of considerable academic interest. Studying abzymes has yielded valuable insights into reaction mechanisms, catalysis, enzyme structure, and function. There is yet the promise of delivering tailored catalysts for difficult reactions for which natural enzymes do not exist. Such catalytic antibodies may be useful even if they do not attain enzyme-like efficiency.

RNA Catalysts (Ribozymes) Substantial role for RNA as information molecule is well established in molecular biology. More recently, RNAs have assumed importance as components of (a) gene silencing through double-stranded small interfering RNAs (siRNA) generated by Dicer, an RNase III-like enzyme, and (b) genome editing through the CRISPR/Cas system where crRNA-guided interference is exploited. The catalytic role of RNA molecules (the so-called ribozymes) has expanded the realm of biological catalysis beyond proteins. RNA catalysts do satisfy the dual criteria of catalysis and specificity but are less impressive catalysts when compared to protein-based enzymes.

They have had much more impact in understanding the origin of life problem and catalytic evolution. Like polypeptides, RNA molecules can fold into higher-order structures that permit the formation of an active site. On the other hand, DNA is predominantly double helical, cannot fold into complicated shapes, and has limited repertoire of chemical groups (it lacks the 20-hydroxyl group) for catalysis. Nevertheless, DNA-based catalysts (DNAzymes) were constructed . More recently, elaboration of different catalytic activities from synthetic genetic polymers (XNAs) was demonstrated . The XNAs (arabino nucleic acids, hexitol nucleic acids, or cyclohexene nucleic acids) fold into defined structures and bind ligands. Few XNAzymes were elaborated directly from random XNA oligomer pools; some of them exhibited in trans RNA endonuclease and ligase activities.

Single-stranded nucleic acid molecules are capable of folding into secondary and tertiary structures. Aptamers are short, single-stranded nucleic acids which bind a variety of ligands with high affinity and specificity. DNA or RNA aptamers can be routinely isolated from synthetic combinatorial nucleic acid libraries by in vitro selection – known as "systematic evolution of ligands by exponential enrichment" (SELEX). Riboswitches are natural versions of aptamers discovered subsequently. The aptamer domains of most riboswitch classes are typically fewer than 100 nucleotides. Riboswitches with distinctive ligand recognition capabilities have been found in all domains of life; they occur with highest frequency within the 50-UTRs of bacterial mRNAs and typically regulate genes involved in metabolism. For instance, glycine riboswitch consists of two different aptamer types that individually bind to a single molecule of glycine.

Cooperative interaction between the two sites allows better sensing of this metabolite. Aptamers and riboswitches are specific ligand binding RNAs with no catalytic function. They may fold or undergo a conformational change upon binding the cognate ligand. An aptamer may be fused to a ribozyme to generate aptazymes. Binding of its cognate ligand to the aptamer displaces it from the stand on which it is bound on the ribozyme, thereby providing for regulation of ribozyme function.

Catalytic RNA was discovered through RNA species capable of auto-cleavage reaction. Subsequently, other activities catalyzed by RNA have been reported.While natural ribozymes catalyze mainly self-cleavage or ligation reactions, they can also accelerate other reaction types (Table 12.3). The RNA component of RNase P is responsible for catalytic processing of precursor tRNAs. The self-splicing group I intron from

Table 12.3 Kinetic constants for some catalytic RNAs

Catalyst	Substrate	k_{cat} (s^{-1})	K_M (M)	k_{cat}/K_M
RNase A	RNA (cleavage; best substrate)	–	–	2.8×10^9
Ribozyme	RNA (cleavage)	2.0×10^{-1}	2.0×10^{-8}	1.0×10^7
Self-splicing intron	RNA (splicing)	1.0×10^{-3}	1.1×10^{-9}	9.0×10^5
Ribosome	(Peptide bond formation)	5.0×10^0	5.0×10^{-3}	1.0×10^3
DNAzyme IV	Ornithine decarboxylase mRNA	1.0×10^{-3}	3.0×10^{-7}	3.0×10^3
Abzyme 34E4 (E2 elimination)	Nitrobenzisoxazole	6.6×10^{-1}	1.2×10^{-4}	5.5×10^3

Catalytic efficiencies of RNase A, a DNAzyme, and an abzyme are also listed for comparison

Tetrahymena thermophila was engineered to perform as a multiple turnover RNA enzyme. The peptide bond formation is attributed to the RNA component of the ribosome .

There are similarities as well as differences between RNA and protein catalysts. Greater structural variety (of amino acids) allows better catalytic properties in protein enzymes than in RNA enzymes. Enzymes are superior catalysts, and many of them function at the diffusion limit. Ribozymes, on the other hand, are rather slow with an apparent maximal rate constant of ~1 min^{-1} .

12.5 ENZYMES MADE TO ORDER

From the general theme of this book and earlier discussion in this chapter , it is obvious that much is known about how enzymes function as catalysts. One measure of how well we understand enzymes is to try and build similar catalysts from first principles. In this sense, de novo enzyme design is an intellectual challenge, and the exercise serves two important objectives.

Table 12.4 Catalytic strategies of protein and RNA enzymes: A comparison

Catalytic strategy	Protein enzyme	RNA enzyme
Substrate orientation and approximation	Yes	Yes
General acid-base catalysis	Always	Deficient[a]
Metal ion catalysis	Many	Always
Organic cofactors	Many	None
Active site electrostatics (dielectric manipulated effectively)	Most proficient	Not proficient
Utilization of binding energy from interactions away from active site	Yes	Yes
Covalent catalysis	Yes	Yes

a. RNA enzymes lack groups with pKas around pH 7.0. The nucleolytic ribozyme GlmS provides an exception to the exclusive use of nucleobases in general acid-base catalysis by ribozymes; a molecule of glucosamine-6-phosphate specifically bound to the RNA structure serves as the general acid in GlmS

It allows experimental validation of the principles of catalysis that we have learned so far. Second, tailor-made catalysts can be built for industrial applications, particularly for those reactions for which natural enzymes do not exist. The design of enzymes with new functions and properties has long been a goal of the protein engineer. However, in enzyme engineering (and de novo design), serendipity continues to outstrip design – a clear sign that our basic understanding of enzyme catalysis to date is far from complete.

Enzyme catalytic power results from a combination of multiple mechanistic strategies . Therefore, valuable insight into the evolution of catalytic function can be gained through de novo design experiments. However, the methods of kinetic analysis discussed earlier in this book will be applicable to all catalysts (synthetic or natural) regardless of their chemical nature. We may note that foundations of enzyme kinetics were laid much before the chemical nature of enzymes as proteins was established by Sumner (using jack bean urease!).

A major benefit of recombinant DNA technology is the ability to do protein and, hence, enzyme engineering. This includes the skill to precisely replace/delete/add one or more amino acids in a given enzyme. These designed yet specific mutations can be engineered on a desired gene (coding for the desired enzyme) with the help of synthetic oligonucleotide constructs as primers . It is very easy to lose an enzyme function through site-directed mutations, but very difficult to gain a new function. Besides point mutations, other functional elements of a protein scaffold could also be replaced/changed through available recombinant DNA tools . The approach is powerful and is anticipated to deliver many tailored enzymes.

Engineering novel enzymes is a rapidly evolving field, and the examples presented here are selective and only representative. A foolproof and robust enzyme activity assay is at the heart of any enzyme engineering and design. While the methodological details are outside the scope of this book, various strategies to generate novel enzyme designs through recombinant DNA technology are outlined below.

Enzyme Redesign One way to generate novel catalysts is to start from an existing enzyme scaffold and rationally alter its structure to effect a predicted change in function. Enzyme redesign may be achieved through simultaneous incorporation and/or adjustment of protein functional elements – through deletion, insertion, loop grafting, and substitution of relevant active site loops, generating chimeras, etc. This could be followed by point mutations to fine-tune the enzyme activity. Using one such approach, a β-lactamase activity was introduced into the scaffold of glyoxalase II. The resulting enzyme completely lost its original activity but catalyzed the hydrolysis of cefotaxime (see Table 12.7 below). Three other interesting examples include the following:

a. The domain swap in bacterial glutamate dehydrogenases to change their pyridine nucleotide specificity .
b. Mixing and matching of different modules of polyketide synthases and non-ribosomal peptide synthases leading to product diversity and the production of hybrid or novel antibiotics .

c. The cancer-associated mutations from isocitrate dehydrogenases which were extrapolated to homologous residues in the active sites of homoisocitrate dehydrogenases, for the catalytic conversion of 2-oxoadipate to (R)-2- hydroxyadipate, a critical step for adipic acid production .

Enzyme redesign as an approach is expected to deliver more tailor-made enzymes in the near future.

The protein engineering approach to redesign a known enzyme has the potential to bring about (a) altered substrate/cofactor specificity and improve catalytic efficiency, (b) enhanced enantioselectivity, (c) a change in metal ion specificity, (d) a desired pH optimum of an enzyme, (e) increased enzyme stability, and (f) alteration of an existing site to catalyze a new chemical reaction. Both close and distant mutations appear similarly effective in improving enzymes in terms of thermostability and catalytic activity. The mutations close to the active site are more effective than distant ones for changing enantioselectivity, substrate selectivity, and alternate catalytic activity of an enzyme . Besides the chemical mechanism first view of enzyme redesign, it may be possible to take the more challenging ligand specificity first approach.

Accordingly, one may introduce catalytic residues into the ligand binding site of a chosen protein to create an active site capable of catalyzing a chemical reaction. After all, the "acid test" of enzyme redesign is to engineer new catalytic activities and aim to change the reaction mechanism itself.

Directed Enzyme Evolution Evolution continues to be an excellent teacher of how to evolve/modify biological catalysts. Directed evolution as a tool accelerates the evolutionary process from millions of years to weeks! And natural enzymes are clearly not evolutionary dead ends. The field of directed evolution of enzymes has made rapid progress and provides a much larger scope and canvas than simple enzyme redesign . It has moved from being a means for studying the relationship between sequence and function to being an extremely powerful tool for optimizing biocatalysts for industry. This Darwinian approach involves generation of random mutant library (generation of genetic diversity) followed by biological selection (of the fittest) for the desired activity. The greatest advantage of directed evolution is its independence from the prior knowledge of enzyme structure. One need not know the nature of interactions between the enzyme and its substrate either.

The procedure of directed evolution consists of (i) generating a random gene library, (ii) expression of these gene variants in a suitable host, and (iii) screening/ selection of library of enzyme forms for the property of interest (Fig. 12.5). It begins with the mutagenesis of the gene encoding the enzyme template of interest (GFE, gene of your favorite enzyme). This important first step should generate a representative yet exhaustive mutant library. The mutated gene library is then inserted into a suitable host (like E. coli or yeast) for expression, and the transformants are plated on selective media. Respective protein variants are chosen from the single colonies growing on these plates. In most directed evolution studies, additional cycles of mutagenesis are necessary for obtaining an optimal catalyst. While many strategies are available (and continue to be developed) for generating exhaustive libraries, the success of a directed evolution effort depends greatly on the method chosen for finding the best mutant enzyme. And, in directed evolution, "we simply get what we screen for"!

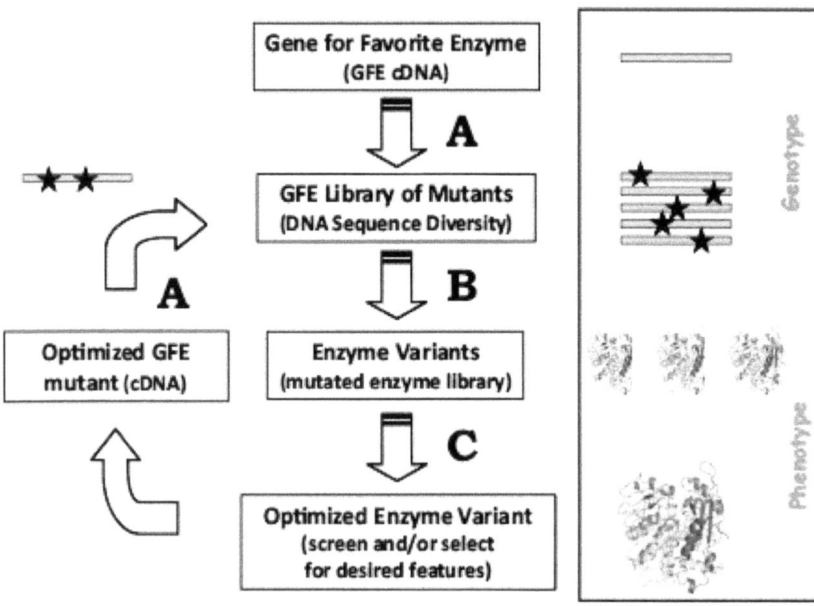

Fig. 12.5 Iterative steps in the directed evolution of an enzyme. Starting with the gene (cDNA) for the favorite enzyme (GFE), a random mutant library is generated (step A). These GFE variants are used to express respective enzyme variants (step B). The optimized variants are screened/ selected (step C) and subjected to further cycles of directed evolution

Table 12.5 Methods to create GFE mutant libraries and generate diversity

Library diversification strategy	Examples and tools
Random mutagenesis	
Chemical and physical mutagenesis	EMS, MNNG, nitrous acid, UV
Error-prone PCR (epPCR)	Supplementing Mn^{2+} and/or unequal dNTP levels
Mutator strains	Mutagenesis plasmid (PACE), XL1-red strain of *E. coli*
Focused mutagenesis	
Site-directed saturation mutagenesis	NNK and NNS codons on mutagenic primers[a]
Computational strategies for high-quality library design	Rosetta design and computationally guided libraries, ISOR, consensus design, REAP, and SCHEMA
Recombination (gene shuffling)	
Homologous recombination	DNA shuffling, domain swapping, family shuffling, StEP, RACHITT, NExT, heritable recombination, ADO and synthetic shuffling
Nonhomologous recombination	ITCHY, SHIPREC, NRR, SISDC and overlap extension PCR (oePCR)

a. Where N can be any of the four nucleotides, K can be G or T, and S can be G or C

Abbreviations and acronyms used in this table include – ADO assembly of designed oligonucleotides, EMS ethyl methanesulfonate, ISOR incorporating synthetic oligonucleotides via gene reassembly, ITCHY incremental truncation for the creation of hybrid enzymes, MNNG N-methyl-N-nitrosoguanidine, NExT nucleotide exchange and excision technology, NRR non-homologous random recombination, PACE phage-assisted continuous evolution, RACHITT random chimeragenesis on transient templates, REAP reconstructed evolutionary adaptive path, SCHEMA a computational algorithm, SHIPREC sequence homology-independent protein recombination, SISDC sequence-independent site-directed chimeragenesis, StEP staggered extension process, UV ultraviolet rays

A range of gene mutagenesis methods are available for generating sequence diversity in a directed evolution experiment. A selection of them is listed in Table 12.5. Choice of a method is critical as each one of them has certain advantages as well as limitations. Features such as biased mutational spectrum, use of hazardous chemicals, uneven sampling of codon space, and need for prior structural, biochemical, or phylogenetic knowledge of each method become important considerations.

Despite these diverse methods to generate gene diversity, it is impossible to cover the entire mutational space available for a typical protein. For a polypeptide chain of "n" amino acids, 20n combinations are possible as there are 20 different amino acids. Even nature has sampled a tiny fraction of these possible sequences over the huge evolutionary timescale. It is therefore best to begin with an existing enzyme sequence. The choice of the sequence diversification method will then depend on the nature of evolutionary trajectory required from the initial scaffold to the desired end point. For instance, directed evolution of enzyme robustness (for stability) may require simultaneous changes scattered across the sequence length.

There are two broad approaches to finding the best mutant enzyme, namely, screening and selection. Screening of the mutant library can be performed in two different ways: (i) facilitated screening wherein mutants are distinguished based on distinct phenotypes and (ii) random screening in which we pick mutants blindly. Selection is always preferred over screening for its higher efficiency. But selection requires a phenotypic functional link between the target gene and its encoding enzyme product that confers the selective advantage (such as better growth, etc.)

Directed evolution has been successfully applied to achieve favorable changes in enzyme properties like stereo- and region-selectivity, expanding the substrate scope and/or activity, enzyme robustness, pH optimum, and promiscuity as catalysts in organic synthesis . Select examples of evolved enzymes reported in the literature are listed in Table 12.6. Industrial-scale biocatalysis applications of optimized enzymes have focused primarily on hydrolases, a few ketoreductases (KREDs), transaminases, oxidative enzymes, aldolases, cofactor regeneration, and enzyme stability in organic solvents . De Novo Enzyme Design Designing an enzyme to catalyze the reaction of one's choice is a grand challenge . This strategy obviously requires detailed knowledge of the protein structure, structural basis of biological catalysis, and computational tools for enzyme design. In principle, the following steps may be envisaged:

- ☆ An appropriate catalytic mechanism is chosen for the target reaction.
- ☆ The transition state for this reaction is described.
- ☆ An idealized active site to position the catalytic residues and maximize TS stabilization be modeled.
- ☆ An appropriate protein scaffold is chosen from the available library.
- ☆ It is optimized in silico to best accommodate the reaction transition state and catalytic residues.
- ☆ The candidate theozyme polypeptides are actually created/produced.
- ☆ They are tested for catalytic activity.
- ☆ Best candidate designer enzyme may be fine-tuned by further sculpting around the transition-state model.

Table 12.6 Examples of directed enzyme evolution

Enzyme	Feature optimized
Monoamine oxidase	Deracemization of racemic amines
Ketoreductases (KREDs)	Chiral intermediates for pharmaceuticals
Laccase	Catalytic efficiency, neutral pH range
Glyphosate N-acetyltransferase	Catalytic efficiency for glyphosate resistance
Cephalosporin acylase	Catalytic efficiency toward adipyl-7-ADCA
ω-Aminotransferase	Specific activity and thermostability
Epoxide hydrolase	Enantioselectivity
Hydantoinase	Inverting enantioselectivity (L-Met process)
β-Lactamase	Antibiotic resistance against cefotaxime
Xylanase, phytase, and lipase	Thermostability
Endoglucanase	Thermal stability, alkaline pH range
Subtilisin	Stability in organic solvent
Halohydrin dehalogenase	Catalytic efficiency
Aldolases	Specificity, efficiency, thermostability
Non-ribosomal peptide synthase–polyketide synthase (NRPS-PKS) hybrid	Produce broad spectrum antibiotic (andrimid)

Although the initial activities of de novo enzyme constructs are typically low, they can be substantially improved through directed evolution approaches. Computational enzyme design has emerged as a promising tool to generate custom-built biocatalysts. The Rosetta de novo enzyme design protocol may be used to tailor enzyme catalysts for a variety of chemical reactions. There are constant efforts to improve the reliability of the Rosetta design cycle . Both the preand post-design analysis of protein structures promises to play an increasingly important role here.

Genetic engineering is relatively easy and routinely accessible now. But to attempt rational enzyme engineering, a vastly improved understanding of protein structure–sequence relationship is required. Our appreciation of protein dynamics is still very limited, and this makes structure–function correlation very hard. Subtle changes in the active site geometry have tremendous unexpected consequences for enzyme function. Hence, rational de novo design of a reasonably efficient enzyme continues to elude us. Few pioneering and brave attempts to build designer enzymes have been made. Relevant examples listed in Table 12.7 highlight the successes and limitations encountered in the de novo design of enzymes.

As can be seen from the representative cases listed above, most designer enzymes have not reached the expected catalytic performance . This is because of our limited mastery over proper protein folding, structure, stability, dynamics, and catalysis. Indeed, subtle changes in the active site geometry are enough to generate remarkable unpredicted consequences for enzyme function. Poor catalytic performance of the de novo designs comes with its own caveats.

Table 12.7 Examples of de novo enzyme design

Designer enzyme	Comments on the design
Oxaloacetate decarboxylase (metal-free)	A rationally designed synthetic 14 amino acid residue cyclic peptide – Oxaldie; decarboxylates oxaloacetate via an imine intermediate on its Lys NH_2. Catalytic efficiency comparable with abzymes (Johnsson et al. 1993)
Dihydrodipicolinate synthase	The N-acetylneuraminate lyase scaffold was rationally redesigned to switch the activity toward dihydrodipicolinate synthase. The designed activity showed 19-fold increased specificity for the new substrate (Joerger et al. 2003)
β-Lactamase (from glyoxalase II)	Several loop grafting steps at the active site to achieve major switch in function. Evolved new activity on an existing glyoxalase II scaffold (Park et al. 2006)
Sesquiterpene synthase	A promiscuous sesquiterpene synthase scaffold was used to build seven novel terpene synthases, catalyzing the synthesis of different sesquiterpenes (Yoshikuni et al. 2006)
Retro-aldolase	Retro-aldol cleavage of a carbon–carbon bond. The mechanism involves enamine catalysis by lysine (via a Schiff base or imine intermediate) giving a catalytic proficiency of 10^4, which is far from natural enzymes (Jiang et al. 2008)
Diels–Alderase (intermolecular)	Organic bimolecular reaction forming two carbon–carbon bonds and up to four new stereogenic centers in one step. Naturally occurring enzymes are not known for this reaction. The designed enzyme is 20 times better than the corresponding abzymes (see Table 39.2) (Siegel et al. 2010)
Triosephosphate isomerase	The Rosetta enzyme design protocol demonstrated for the triose phosphate isomerase reaction as an example (Richter et al. 2011)
Kemp eliminase	A well-studied organic model system for proton transfer from carbon. In Kemp elimination, the deprotonation of substrate (5-nitrobenzisoxazole) leads to electronic rearrangements that break the C–H and N–O bonds while forming a C≡N triple bond. The designer enzyme accelerates the elementary chemical reaction (6×10^8 fold), nearly as efficient as natural enzyme like triose phosphate isomerase (Blomberg et al. 2013)

Ensuring that the observed activity is really due to the designed enzyme, and not a contaminating activity from the expression host, is critical. Attempts to convert catalytically inert ribose-binding protein into an active triose phosphate isomerase ran into such difficulties and elicited the response – "It is a bush-league error not to purify your proteins well, especially in such work".

De novo protein design allows us to explore the full sequence space. Computational methodology has progressed well for a wide range of structures to be designed from scratch and with atomic-level accuracy . However, obtaining more active catalysts will require improved control over substrate binding and better pre-organization of the active site. Modifying existing protein scaffolds through rational redesign has been a more fruitful option so far – where the emphasis on finding what works rather than predicting what works .

The fine-tuning of engineered enzymes can only be fulfilled today by combinatorial approaches. The marriage of rational design and directed evolution (combinatorial redesign!) seems to be the way to go at present.

REFERENCES

1. Branden, C., and Tooze, J. (1991) *Introduction to Protein Structure*, Garland, New York.
2. Chotia, C. (1975) *J. Mol. Biol.* 105, 1.
3. Copeland, R. A. (1994) *Methods for Protein Analysis:A Practical Guide to L aboratory Protocols*, Chapman & Hall, New York.
4. Creighton, T. E. (1984) *Proteins, Structure and Molecular Properties*, Freeman, New York.
5. Davis, J. P., and Copeland, R. A. (1996) Protein engineering, in *Kirk-Othmer Encyclopedia of Chemical Technology*, Vol. 20, 4th ed., Wiley, New York.
6. Dayoff, M. O. (1978) Atlas of Protein Sequence and Structure, Vol. 5, Suppl. 3, National Biomedical Res. Foundation, Washington, D.C.
7. Dill, K. A., Bromberg, S., Yue, K., Fiebig, K. M., Yee, D. P., Thomas, P. D., and Chan, H. S. (1995) *Protein Sci.*, 4, 561.
8. Dixon, M., and Webb, E. C. (1979) *Enzymes*, 3rd ed., Academic Press, New York.
9. Dugas, H., and Penny, C. (1981) *Bioorganic Chemistry, A Chemical Approach to Enzyme Action*, Springer-Verlag, New York.
10. Hansch, C., and Coats, E. (1970) *J. Pharm. Sci.* 59, 731.
11. Klapper, M. H. (1977) *Biochem. Biophys. Res. Commun.* 78, 1018.
12. Kyte, J., and Doolittle, R. F. (1982) *J. Mol. Biol.* 157, 105.
13. Pauling, L., Itano, H. A., Singer, S. J., and Wells, I. C. (1949) *Science*, 110, 543.
14. Picot, D., Loll, P., and Garavito, R. M. (1994) *Nature*, 367, 243.
15. Stryer, L. (1989) *Molecular Design of L ife*, Freeman, New York.
16. Walsh, C. (1979) *Enzyme Reaction Mechanisms*, Freeman, New York.

INDEX

www.ingramcontent.com/pod-product-compliance
Ingram Content Group UK Ltd.
Pitfield, Milton Keynes, MK11 3LW, UK
UKHW021328190726
13845UKWH00004B/72

9 789349 964167